**DO NOT REMOVE
CARDS FROM POCKET**

SYNTHETIC RUBBERS
Processes and Economic Data

FROM THE SAME PUBLISHER

Publications in English

- Characterization of Heavy Crude Oils and Petroleum Residues.
  Caractérisation des huiles lourdes et des résidus pétroliers.
  International Symposium, Lyon 1984.

- VIIth International Symposium on Alcohol Fuels. VIIe Symposium
  international sur les carburants alcoolisés. Paris 1986.

- Hydrocarbon Chemistry of FCC Naphtha Formation (The). Proceedings
  of the Symposium of the Division of Petroleum Chemistry, Inc., American
  Chemical Society, Miami 1989.
     Edited by H.J. LOVINK and L.A. PINE

- Petrochemical Processes. Technical and Economic Characteristics.
     A. CHAUVEL, G. LEFEBVRE
     Volume 1. Synthesis-Gas Derivatives and Major Hydrocarbons.
     Volume 2. Major Oxygenated, Chlorinated and Nitrated Derivatives.

- Catalytic Cracking of Heavy Petroleum Fractions.
     D. DECROOCQ

- Applied Heterogeneous Catalysis. Design. Manufacture. Use of Solid Catalysts.
     J. F. LE PAGE

- Chemical Reactors. Design. Engineering. Operation.
     P. TRAMBOUZE, H. VAN LANDEGHEM, J.P. WAUQUIER

- Methanol and Carbonylation.
     J. GAUTIER-LAFAYE, R. PERRON

- Commodity Thermoplastics. Technical and Economic Characteristics.
     J.P. ARLIE

- Resid and Heavy Oil Processing
     J.F. LE PAGE, S. G. CHATILA, M. DAVIDSON

INSTITUT FRANÇAIS DU PÉTROLE

ÉCOLE NATIONALE SUPÉRIEURE DU PÉTROLE ET DES MOTEURS

Jean-Pierre ARLIE

Chief Engineer

Institut Français du Pétrole

SYNTHETIC RUBBERS

Processes and Economic Data

Translation from the French
by Nissim Marshall

Distributed in the United States by
GULF PUBLISHING COMPANY
Houston, Texas

1992

ÉDITIONS TECHNIP 27 RUE GINOUX 75737 PARIS CEDEX 15 techniP

Translation (reviewed) of
"Caoutchoucs synthétiques. Procédés
et données économiques". J.P. Arlie
© Editions Technip, Paris 1980

ISBN 2-7108-0619-3

Printed in France
by Imprimerie CHIRAT, 42540 St-Just-la-Pendue

PREFACE

Specialists from the petroleum industry, from the *Institut Français du Pétrole (IFP)* and the *Ecole Nationale Supérieure du Pétrole et des Moteurs (ENSPM)* and from universities teach at the *Graduate Study Center for Refining and Chemical Engineering of ENSPM.*

In conjunction with this teaching, they have written various books for educational purposes, dealing with the different scientific and technical aspects of these petroleum operations. The present book is one of them.

M. Verwaerde

CONTENTS

Chapter 1
SYNTHETIC RUBBERS

Chapter 2
STYRENE/BUTADIENE RUBBER (SBR)

J.P. ARLIE

CONTENTS

Chapter 3
POLYBUTADIENE

Chapter 4
POLYISOPRENE

Chapter 5
ETHYLENE/PROPYLENE RUBBERS

J.P. ARLIE

CONTENTS

Chapter 6
BUTYL RUBBER

Chapter 7
POLYCHLOROPRENE RUBBER

CONTENTS

Chapter 8
NITRILE RUBBER

Chapter 9
THERMOPLASTIC ELASTOMERS

CONTENTS

J.P. ARLIE

INTRODUCTION

The world synthetic rubber industry now produces more than 10 million tons annually, ranking it second among the major petrochemical industries, behind the plastics industry.

The production of synthetic elastomers, which was virtually non-existent in 1950, has grown rapidly since then, in competition with that of natural rubber from *Hevea brasiliensis*. Figures on world demand in 1989 show that synthetic elastomers account for two-thirds of total consumption, with wide variations in this ratio from one part of the world to another, and with the highest ratios in the industrialized countries.

The various reasons underlying the rapid growth of synthetic rubbers are both economic and strategic: economic, because natural rubber is only produced in the developing countries, with a tropical climate. Since their output is insufficient to satisfy market demand, the strategic need arose to avoid a supply shortage for certain countries, as well as wide swings and sharp rises in the price of natural rubber.

The major economic powers that showed an early interest in the production of synthetic rubber initially obtained a good replica of natural rubber. Subsequently, a number of mechanical properties which proved to be inadequate were improved, and several types of synthetic rubber thus appeared, normally classed into two main categories.

The first category concerns **general-purpose rubbers**, and is the largest in terms of tonnage and value. It corresponds to any elastomer, used alone or as the major component of a blend, and applies to all uses. Pneumatic tires represent the largest application (60%), involving the following elastomers: **styrene butadiene copolymers (SBR), polyisoprene, polybutadiene, and ethylene/propylene copolymers.**

All elastomers not principally used to manufacture tires are called specialty rubbers. They are employed in areas demanding specific performance, such as resistance to chemicals. The leading examples are **butyl rubber, polychloroprene, nitrile rubber**, and **thermoplastic elastomers**.

For each product examined, we shall try to identify the principal features concerning:

(a) The main mechanisms of polymerization chemistry.
(b) Feedstock purity.
(c) Polymerization processes:

  · Continuous or batch.
  · Reaction medium.
  · Type of reactor.

(d) Polymer recovery.
(e) Finishing operations.
(f) The scale factor of the process.

We shall also discuss:

(a) Technical and economic characteristics.
(b) Battery limits investments: the figures given concern plants built in Western Europe in 1979, and they correspond to averages.
(c) Material balance and energy consumption.
(d) Main uses in the leading consumer countries.
(e) Installed capacities in various producing countries.

Chapter **1**

SYNTHETIC RUBBERS

Although synthetic elastomers use a number of monomers identical to those of plastics for their synthesis, such as ethylene, propylene, butadiene and styrene, they are different in their physical properties, synthesis processes and transformations.

The main qualities desired in an elastomer are high elasticity, high tensile strength, low energy dissipation by hysteresis, good abrasion resistance, good ageing behavior, and good resistance to chemical degradation.

Any polymer substance which, after being stretched to at least twice its initial dimension, can rapidly recover its initial size after the release of the stretching force, satisfies the definition of an elastomer. However, very few polymers, if any, actually correspond to this definition in the raw state. The macromolecules which form a viscous mass slide over each other under the effect of a stretching force. If this force is removed, the substance in fact does not return to its initial state. The insoluble three-dimensional network which meets the definition of the rubbery state is obtained by a chemical reaction: vulcanization. This operation is generally conducted by the addition of chemical substances such as sulfur. Since most synthetic elastomers are unsaturated compounds, the sulfur is added to the α of the double bond, but the unsaturation is not eliminated. This change is irreversible, and this characteristic is related to the transformation undergone by thermosetting polymers.

1.1 HISTORICAL BACKGROUND

The first industrial synthetic elastomer was made in Germany during the First World War. This polydimethylbutadiene (methyl rubber) was not competitive with natural rubber either technically or economically. Its production was discontinued when the War ended. In 1926, the first commercial elastomer introduced by *I.G. Farben* appeared in Germany. It was called Buna and was obtained by the polymerization of butadiene catalyzed by sodium. Despite the difficulties of its application and its poor mechanical properties for use in pneumatic tires, this elastomer continued to be produced in Germany until 1945 and in the USSR until 1950.

The difficulties of obtaining isoprene, which had been isolated from natural rubber in 1860, diverted research and development work to butadiene, which polymerized with styrene to give Buna S. In 1935 *I.G. Farben* started the industrial manufacture of acrylonitrile/butadiene copolymers. War preparations accelerated the manufacture of Buna S, and its production reached 126,000 tons in Germany in 1943.

The United States developed other elastomers in the 1930s: neoprene by *Du Pont*, and butyl rubber made by *Standard Oil*, obtained by the copolymerization of isobutene and small amounts of butadiene with the aid of ionic initiators. However, the principal elastomer manufactured was GRS, a styrene/butadiene copolymer of which 800,000 tons were produced in 1945. GRS became SBR, which continued to grow steadily from 1955 on.

These polymers displayed no characteristic of stereoregularity, whereas natural rubber, which is made up of a chain of cis-1,4 isoprene units, is a totally different compound from gutta percha, which is made up of a chain of trans-1,4 units. Polybutadiene, polyisoprene and ethylene/propylene copolymers only appeared in 1960, with the applications of stereospecific polymerization due to the work of K. Ziegler and G. Natta. Table 1.1 summarizes the industrial production of synthetic elastomers.

TABLE 1.1

PRODUCTION OF PRINCIPAL SYNTHETIC RUBBERS
STARTUP DATE AND LOCATION

Chemical name	Trade Name	Starting date	Company	Country
Copolymer of butadiene with styrene or acrylonitrile	Buna S or N	1930/1940	*I.G. Farben*	Germany
Polychloroprene	Neoprene	1930/1940	*Du Pont*	USA
Isobutene/isoprene copolymer	Butyl rubber	1943	*Esso*	USA
Styrene/butadiene copolymer	GRS then SBR	1941/1942	State	USA
Styrene/butadiene copolymer	SK	Around 1940	State	USSR
Polybutadiene	BR	1961	*Goodyear Shell*	USA
Polyisoprene	IR	1962	*Shell Goodyear*	USA
Ethylene/propylene copolymer	EPDM	1963	*Du Pont Montecatini*	USA Italy

1.2 GENERAL-PURPOSE RUBBERS

The most important application of these elastomers is the manufacture of pneumatic tires. Synthetic rubber is used either pure, blended with natural rubber, or in a blend with other synthetic rubbers. These blends are governed by the mechanical properties desired in the tires.

For this application, the elastomers are nearly always extended with oil to minimize internal heating. Variable amounts of carbon black are added to improve the hardness of the rubber and the road characteristics of the tire.

The usable polymers are derived from monomers obtained by the steam-cracking of naphtha. The common feature of most of these elastomers is the residual double bond which favorizes vulcanization.

Statistical styrene/butadiene copolymer (75% butadiene by weight) is mainly used in treads of light automotive vehicles, alone or blended with natural rubber.

Polybutadiene gives the tires high abrasion resistance, excellent low-temperature strength (the best of the general-purpose rubbers) and very good ageing behavior. However, it exhibits poor adherence to a wet surface, giving rise to aquaplaning. This is why it is employed in a blend with SBR or natural rubber.

cis-1,4 polyisoprene is a nearly perfect replica of natural rubber, and can hence be substituted for it without any difficulty.

The rubbers obtained by **ethylene/propylene copolymerization** and denoted EP are incompatible with the other general-purpose elastomers. They are also difficult to vulcanize and lack tackiness. The introduction of a termonomer (hexadiene, dicyclopentadiene) allows vulcanization by the usual sulfur processes, but significantly raises its cost.

1.3 SPECIALTY RUBBERS

World demand for specialty rubbers is about 10% of total rubber demand. Many elastomers fall into this category, and the chief ones are discussed below.

Butyl rubber is a copolymer of isobutene with about 2% isoprene. It is mainly used in the production of inner tubes due to its excellent impermeability to gases.

Polychloroprene or neoprene is used in a wide variety of applications related to its resistance to oils and solvents.

J.P. ARLIE

Nitrile rubber, a butadiene/acrylonitrile copolymer (75% butadiene by weight) is mainly used for its excellent resistance to oils and aromatic solvents. However, its processing is difficult.

Thermoplastic elastomers are trisequence styrene/butadiene/styrene or styrene/isoprene/styrene copolymers. These materials display properties of elastomers, but can be processed like thermoplastics without vulcanization.

1.4 POLYMERIZATION METHODS

Two methods are mainly used, emulsion polymerization initiated by free radicals, and solution polymerization in the presence of ionic initiators or catalysts.

1.4.1 Emulsion Polymerization by Free Radical Reaction

These processes were developed to reproduce by synthesis the biological conditions of the formation of natural rubber latex. The preparation of latex with soaps and emulsifiers and the development of water-soluble free radical promoters favored this technique, from the time when high molecular weight polymers were obtained with high polymerization rates. Emulsion polymerization has the following advantages:

(a) Excellent transfer of the heat of reaction by means of water.
(b) Lower viscosity, at equivalent molecular weight, than any homogeneous system.

The drawback of this type of polymerization is that it can only take place in the presence of free radical initiators.

1.4.2 Solution Polymerization by Ionic Reaction

Stereoregular elastomers, which cannot be obtained by the free radical method, are manufactured by processes in which polymerization takes place in solution in an inert solvent (hydrocarbons) to lower the viscosity and the molecular weight.

The systems which initiate or catalyze the reaction are the key to the properties of the polymers obtained.

1.4.2.1 Anionic Polymerization

The chains are initiated with the aid of lithium compounds. In the absence of impurities, each lithium atom initiates polymerization. The polymer is active and the propagation reaction continues until the monomer is totally depleted. In contrast to free radical polymerization, no chain transfer or short-stop reaction occurs. The reaction is hence terminated by deactivation of the chain by means of polar compounds (water, alcohols).

Since the initiation rate is faster than the propagation rate, the resulting molecular weight distribution curve is much narrower than in emulsion polymerization ($\overline{M_w}/\overline{M_n}$ = 1.5 against 4). Consumption of initiator is low.

These initiators help to polymerize dienes and to obtain statistical and sequenced copolymers.

1.4.2.2 Polymerization by Coordination Catalysts

Derived from the work of Ziegler and Natta, coordination catalysts are formed by the interaction of organic metallic compounds of groups I and III of the periodic table of elements, with halogenides and other derivatives of the transition metals of groups IV to VIII:

(a) The organometallic reducing element is of the type $Al(C_2H_5)_3$ or $Al(C_2H_5)_2 Cl$.
(b) The transition metallic compound may be Ti, Co or V.

A large number of theoretical models has been suggested for the stereospecific polymerization mechanisms with the aid of these complexes, but none of them accurately depicts the experimental reality. It is now acknowledged that an organometallic complex is formed at the onset of the reaction, providing the active site of polymerization.

These catalysts are capable of polymerizing butadiene and isoprene, and of copolymerizing ethylene with propylene, because they permit stereochemical control of the growing chain.

1.4.2.3 Cationic Polymerization

In a cationic polymerization system, the initiators consist of strong acids H_2SO_4 or $HCl\ O_4$, or Lewis acids like BF_3 or $AlCl_3$, which give rise to carbonium ions. The carbonium ions are attached to the double bond. The reaction is faster at low temperature. Since the carbonium ions are highly reactive, transfer reactions are frequently observed. This type of polymerization is used to produce butyl rubber.

Chapter **2**

STYRENE/BUTADIENE RUBBER (SBR)

Rubbers based on styrene and butadiene, which were developed from 1945 in the United States and Germany, today account for over 50% of world synthetic rubber production.

The respective proportions of butadiene and styrene in the copolymer are about 75 and 25% by weight.

The polymerization processes are continuous, by free radical emulsion, and anionic solution polymerization.

2.1 CHEMISTRY OF STYRENE/BUTADIENE COPOLYMERIZATION

The copolymerization reaction can be represented roughly as follows:

$$n \left[C_6H_5 - CH = CH_2 + CH_2 = CH - CH = CH_2 \right] \longrightarrow$$

$$\left[\begin{matrix} CH - CH_2 - CH_2 - CH = CH - CH_2 \\ | \\ C_6H_5 \end{matrix} \right]_n$$

$\Delta H^{\circ}_{298} = 3.9$ kcal/mol.

2.1.1 Emulsion Copolymerization

The monomers, initiator, salts, soap and chain transfer agents are dissolved in water. The polymerization reaction begins with the monomers, which are dissolved in the soap micelles 50 Å in diameter. The radicals formed from the initiator are captured by these monomers. The growing polymer particles absorb the soap from the micelles, which are destroyed when conversion reaches 10 to 15%. The polymer continues to grow by absorption of the monomer and of the free radicals in the particles 200 Å in diameter.

The following promoters are used in this type of copolymerization:

(a) Potassium persulfate combined with a mercaptan. The latter plays a dual role: creation of free radicals and limitation of molecular weights by chain transfer reaction. The reactions are as follows:

$$K_2S_2O_8 + 2RSH \rightarrow 2KHSO_4 + 2RS. \text{ (initiation)}$$
$$RS. + M \rightarrow RSM.$$
$$RSM. + nM \rightarrow RSM_{(n+1)}. \text{ (propagation)}$$
$$RSM_{(n+1)}. + RSH \rightarrow RSM_{(n+1)}H + RS. \text{ (chain transfer)}$$
$$R. + R. \rightarrow RR \text{ (termination by radical combination)}$$
$$RCH_2 - CH_2. + R. \rightarrow RH + RCH = CH_2 \text{ (termination by dismutation)}.$$

(b) Redox promoters for reactions at low temperature (+5°C). They include sodium sulfoxylate/formaldehyde as an oxidizing agent, a hydroperoxide as an oxidant, an iron salt and a chelatant.

2.1.1.1 Monomer Purity

The purity specifications required are listed in Tables 2.1 and 2.2.

TABLE 2.1
STYRENE SPECIFICATIONS FOR EMULSION SBR

Compound	% by weight maximum
Styrene	99.000 (% by weight minimum)
Polystyrene	0.005
Benzaldehyde	0.030
Peroxides	0.010
Sulfur	0.005
Chlorides	0.010
Paratertbutylcatechol[1]	10 to 15 ppm

1. Added as anti-oxidant.

TABLE 2.2
BUTADIENE SPECIFICATIONS FOR EMULSION SBR

Compound	Specification
Butadiene	99% by weight minimum
Dimers	0.1% by weight maximum
C_5 and C_5+	0.4% by weight maximum
Non-volatile residues	0.1% by weight maximum
Oxygenated compounds	0.3% by volume maximum
Tertbutylcatechol	100 ppm
Peroxides expressed as H_2O_2	10 ppm
Acetylenics	350 ppm
1,2-butadiene	100 ppm
Carbonyl compounds	600 ppm
Sulfur	5 ppm
Propadiene	50 ppm

2.1.1.2 Molecular Weight

The weight average molecular weight ranges from 2 to $4 . 10^5$, which is much lower than that of natural rubber ($1.5 . 10^6$). The polydispersity defined by the ratio of the weight average molecular weight to the number average molecular weight ranges between 4 and 6, against 3 for natural rubber.

2.1.1.3 Molecular Configuration

As we shall show in greater detail in the chapter on the polymerization of butadiene, its configurations in a polymer chain are cis-1,4, trans-1,4 or vinyl 1,2. The distribution of these three configurations in the styrene/butadiene copolymers obtained by the free radical method essentially depends on the polymerization temperature, as shown by the following data:

Reaction temperature (°C)	Percentage by weight		
	cis-1,4	trans-1,4	1,2
− 18	73.4	4.6	22.0
5	68.6	8.0	23.4
50	58.2	14.9	26.9

2.1.2 Solution Copolymerization

Industrial processes operate by the anionic method using an alkyl lithium as initiator, which is generally butyl lithium. This type of copolymerization yields SBR copolymers consisting of unbranched linear chains.

One of the characteristics of the copolymerization of butadiene and styrene is the difference between their respective reactivities in non-polar solvents such as hydrocarbons. At the onset of the reaction, the butadiene polymerization rate is greater than that of styrene, whereas the opposite occurs at the end of the reaction. To obtain a statistical copolymer in which the styrene and butadiene units are distributed at random in the polymer chains, the reactivity ratios are altered by adding a polar solvent in small proportions (< 5%) to the hydrocarbon solvent.

2.1.2.1 Monomer Purity

The purity specifications for styrene are approximately the same as in the emulsion process (Table 2.1). They are slightly more severe for butadiene (Table 2.3).

TABLE 2.3

BUTADIENE SPECIFICATIONS FOR STYRENE/BUTADIENE
SOLUTION COPOLYMERIZATION

Compound	Specification
Butadiene	99% by weight minimum
C_5 and C_5+	0.4% by weight maximum
Dimers	0.1% by weight maximum
Non-volatile residues	0.1% by weight maximum
Oxygenated compounds	0.3% by volume maximum
Tertbutylcatechol	100 ppm
Peroxides expressed as H_2O_2	5 ppm
Acetylenics	100 ppm
Methyl acetylene	10 ppm
Ethyl acetylene	50 ppm
1,2-butadiene	150 ppm
Carbonyl compounds	50 ppm
Sulfur	5 ppm
Methanol	15 ppm
Propadiene	50 ppm

2.1.2.2 Molecular Weight

The molecular weight is theoretically 100/(BuLi) considering that (BuLi) is the quantity, expressed in gram molecules, consumed per 100 g of monomer. If, for example, we take a number average molecular weight of 300,000, the amount of butyl lithium required is 0.03 kg per 100 kg of monomer, or about 0.3 kg/t of SBR.

Given the fact that the initiation rate is faster than the propagation rate, the polydispersity is significantly lower than that of SBRs obtained by emulsion copolymerization.

2.1.2.3 Molecular Configuration

The butadiene configurations depend on the polarity of the solvent and also on the copolymerization temperature. In a non-polar solvent, the configuration varies from 35 to 55% of trans-1,4 units, 38 to 60% of cis-1,4 units, and 5 to 8% for 1,2 units.

The addition of small amounts of a polar solvent boosts the percentage of 1,2 units, which may rise to 70%.

2.2 SBR SYNTHESIS PROCESSES

SBR is manufactured by two types of industrial process:

(a) Processes in which polymerization is carried out by free radicals in emulsion in water and at low temperature (cold emulsion polymerization). Note that the hot polymerization method (hot rubber) which used persulfates as initiators, was discarded in favor of cold polymerization, which spread with the adoption of redox systems.
(b) Anionic solution polymerization processes.

2.2.1 Cold Emulsion Processes (Fig. 2.1)

This is the most widely-used technique, and accounts for 90% of world production capacity. All the processes are continuous and generally highly automated. They are equipped to produce many types of SBR.

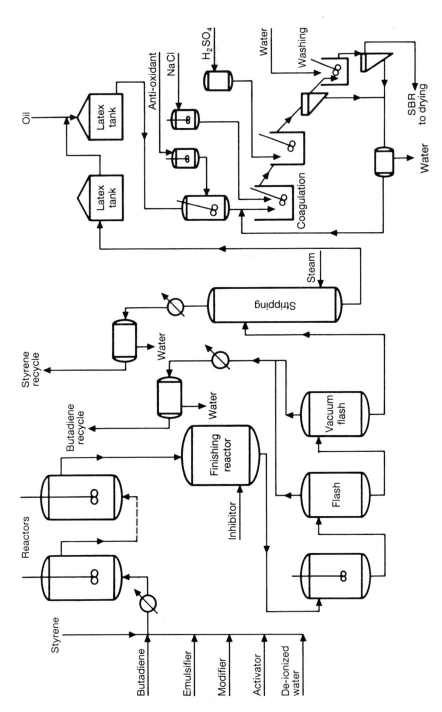

Fig. 2.1 Styrene/butadiene cold emulsion polymerization process.

The process licensors are *Firestone Tire and Rubber Company* (USA), *Goodrich* (USA), *Goodyear* (USA), *Polymer Corporation* (Canada), and *International Synthetic Rubber* (United Kingdom).

Every installation has four sections:

(a) Reactant preparation.
(b) Polymerization.
(c) Monomer recovery.
(d) Rubber coagulation and drying.

Several industrial plants have been built with capacities of 35,000 t/line. It appears possible to build units up to 70,000 t/year in a single line.

2.2.1.1 Reactant Preparation

The monomers are treated with caustic soda in stirred tanks to remove the polymerization inhibitors used for monomer storage and transport. The effluents are then washed with water to remove any remaining caustic. The two monomers, part of which represents the recycle stream after the reaction, are mixed in butadiene/styrene weight proportions of 3 to 1.

Weight and make-up tanks are used to prepare the different emulsions and solutions required for the reaction section or the product finishing section.

A. Soap Solution

This is used as a feed emulsifier. Its composition depends on the type of end product desired. It is usually a solution of a fatty acid soap or carboxylic acid salts, such as versatic acid or benzoic acid.

B. Initiator

All the processes use redox systems. The reducing agent is often sodium sulfoxylate. The oxidizing agent is either cumene hydroperoxide or, preferably, paramenthane hydroperoxide, which allows for faster reaction rates, given its aptitude to rapid decomposition. Ferrous sulfate is the chelatant.

C. Shortstopping

In the overwhelming majority of cases, monomer conversion is less than 65%, because higher conversion causes partial gelling of the polymer. To guarantee uniform product quality, the reaction is stopped as soon as the desired conversion is obtained. Various inhibitors in solution are used, such as sodium dimethyldithiocarbamate.

D. Stabilizers

These are emulsions that are added to the latex before coagulation to prevent degradation by oxidation and polymer cross-linking during the finishing and storage operations. Various stabilizers are used, including N-phenyl β naphthylamine (Neozone D®, PBNA®, 2246® or Ac-5®).

E. Coagulants

Polymerization gives rise to a latex, i.e. a viscous mass in emulsion. If the solid elastomer is desired, the latex must be coagulated by the addition of chemicals. The chief coagulant is sulfuric acid containing sodium chloride solution.

F. Molecular Weight Regulators

The molecular weight of the final product is regulated by mercaptans like dodecyl mercaptan, which helps to limit the molecular weight by causing chain transfers.

2.2.1.2 Polymerization Reaction

A typical production formula is shown in Table 2.4.

The reaction takes place in a series of stirred reactors, **at a temperature of 5°C and 1 to 4 bar pressure to keep the butadiene in the liquid state.** Polymerization time is about 10 h.

TABLE 2.4

TYPICAL PRODUCTION RECIPE OF SBR 1500

Product	Parts by weight
Butadiene (98% +)	72.00
Styrene (98% +)	28.00
Water	180.00
Fatty acid soap	4.50
Other surfactant	0.30
Dodecyl mercaptan	0.20
P-menthane hydroperoxide	0.63
Ferrous sulfate	0.01
Sodium sulfoxylate	0.05

Each reactor, with a capacity of 15 to 20 m^3, is kept under inert atmosphere to prevent any cross-linking. These reactors are jacketed, and equipped with a cold brine (ammonia) circulating pump. An installation with a production capacity of 40,000 t/year of dry polymer requires ten reactors in series.

The emulsion passes through each reactor in upflow for 1 h before going to the next reactor. For total conversion of 60%, monomer conversion per reactor is hence 6%.

A solution of dodecyl mercaptan is introduced into the final reactor to stop polymerization. An additive such as hydrazine or a derivative of hydroxylamine is used to prevent the formation of foam (popcorn) when the latex is heated.

The latex is pumped to a buffer tank kept under 4 bar pressure at 50°C by open steam injection.

2.2.1.3 Monomer Recovery

The unreacted 40% of the monomers must be recovered and then recycled.

The butadiene is vaporized in two flash tanks in series. The last traces of butadiene are removed by means of a vacuum pump. It is cooled, recompressed, and then sent to a settler, where it separates from the water. It is then pumped to a storage tank in the presence of an inhibitor.

The butadiene-free latex is pumped to a tray column at the bottom of which steam is injected (5 bar) to strip the styrene monomer. This is cooled and sent to a settling tank, where it separates from the entrained water. It is then pumped to the storage tank.

2.2.1.4 Coagulation and Drying

The latex leaving the column bottom is cooled and then stored in homogenization tanks (unit volume 800 m^3). The number of these tanks depends on the range of SBR grades that the unit has to produce (generally between three and six). The anti-oxidant N-phenyl β naphthylamine (about 1% by weight) is added to the latex, which is then coagulated by the successive addition of salt and dilute sulfuric acid. By breaking the emulsion, the acid allows the copolymer to precipitate in the form of crumbs, which are washed with water to remove inorganic impurities.

The polymer, which contains about 50% water, is then dried (tunnel furnace) and pressed into 40 kg bales.

2.2.2 Solution Polymerization Processes

These processes represent 10% of world capacity, and are used in some countries in addition to the emulsion process. The solution method offers the advantage of great flexibility, because it allows the production of SBR or polybutadiene by using lithium-based polymerization initiators. However, SBR grades from solution processes are more difficult to process than polymers from emulsion processes, hindering their use for pneumatic tires.

The process licensors are *Firestone Tire and Rubber Company* (USA), *Phillips Petroleum Company* (USA) and *Shell* (Netherlands).

Capacities per line range from 25,000 to 30,000 t/year.

These processes, which are very similar to the butadiene solution polymerization process described in Chapter 3, are suitable for capacities up to 100,000 t/year.

A number of salient points is worth noting:

(a) The initiator is butyl lithium.
(b) The solvent is a hydrocarbon such as hexane. The solvent/monomer weight ratio is 8. This gives a more viscous polymer at the end of the reaction, while ensuring good stirring of the reactor and good heat transfer. A higher monomer concentration in the solvent would help to increase the polymerization rate and to decrease the number of reactors, but would demand a larger heat exchange area and would limit the molecular weight of the monomer, given the higher viscosity of the reaction medium.
(c) The reactors are of vitrified steel, jacketed, and equipped with a turbine stirrer.
(d) The polymerization reaction takes place at 1.5 bar and 50°C. Reaction time is 4 h for 98% conversion.
(e) The purification system by flash and steam stripping should serve to achieve maximum hexane recovery, and also to concentrate the SBR from the slurry (the slurry is the concentrated 10 to 15% polymer solution).
(f) The finishing operations are the same as those described for the emulsion process.

2.3 CHARACTERISTICS OF SBR RUBBERS

According to the code of the *International Institute of Synthetic Rubber Producers (IISRP)*, SBR copolymers are classed in different categories:

(a) **SBR series 1000:**
   Copolymers obtained by hot polymerization.

(b) **SBR series 1500:**
   Copolymers obtained by cold polymerization. Their properties depend on the reaction temperature and on the styrene and emulsifier content. The variation in these parameters affects the molecular weight and hence the properties of the vulcanized mixture.

(c) **SBR series 1700:**
   SBR 1500 extended with oil.

(d) **SBR series 1600 and 1800:**
   Carbon black is mixed with SBR 1500 gum during production by the incorporation of an aqueous dispersion of carbon black with the SBR latex previously extended with oil. A master mixture approaching the finished product is obtained after coagulation and drying.

2.4 ECONOMIC DATA

Table 2.5 provides a glance at the investments and various consumption figures concerning SBR rubber. Given the number of grades of SBR, the figures represent averages.

2.5 USES AND PRODUCERS

A comparison of SBR with natural rubber reveals the following differences:

TABLE 2.5

ECONOMIC DATA FOR SBR PRODUCTION

Process	Emulsion		Solution
Capacity (t/year)	100,000 (50% type 1500) (50% type 1700)		100,000
Battery limits investments 10^6 FF 1989	400		430
Consumption per ton of SBR	Type 1500[1]	Type 1700[2]	Type 1500
Raw materials:			
· Butadiene (t)[3]	0.720	0.524	0.760
· Styrene (t)	0.220	0.160	0.253
· Aromatic oil (t)	–	0.275	–
Chemicals and catalysts:			
· Soap (kg)[4]	70.0		
· Sodium chloride (kg)	200.0		
· Sulfuric acid (kg)	24.0		
· Miscellaneous chemicals (kg)	25.0		
· N-butyl lithium (kg)			1.0
· N-hexane (kg)			40.0
· Stearic acid (kg)			5.0
· Stabilizer (kg)			10.0
· Tetrahydrofuran (kg)			0.5
Utilities:			
· Electricity (kWh)	460.0		490.0
· Steam (t)	3.5		4·6
· Process water (m^3)	4·0		–
· Cooling water (m^3)	180.0		210·0
Labor (operators per shift)	15.0		11.0

1. **Type 1500 SBR**: not extended with oil and not pigmented, cold polymerized continuously.
2. **Type 1700 SBR**: oil extended, pigmented, cold polymerized continuously.
3. **(t) = metric ton.**
4. **(kg) = kilogram.**

2. STYRENE/BUTADIENE RUBBER (SBR)

(a) SBR is inferior to natural rubber for processing, tensile and tear strength, tack and internal heating.
(b) SBR is superior for permeability, ageing, and resistance to heat and wear.
(c) The vulcanization of SBR requires less sulfur, but more accelerator.
(d) The reinforcing effect of carbon black is much more pronounced on SBR than on natural rubber.
(e) For use in tires, SBR is better for passenger vehicles, whereas natural rubber is preferable for utility vehicles and buses.
(f) Oil-extended SBRs are primarily used for manufacturing tires, belting and shoe soles. SBR master blends are employed in the mass production of tire treads.

The main uses of SBR are listed by percentage in Table 2.6.

TABLE 2.6
USES OF SBR IN THE UNITED STATES IN 1989

Use		%
Tires		69
· Passenger vehicles	49	
· Utility vehicles and buses	10	
· Other tires	4	
· Tire treads	6	
Non-tire uses		31
· Automotive applications	9	
· Mechanical objects	20	
· Other	2	
Total		100

Production capacities in 1989 for a number of producing countries are listed in Table 2.7.

2. STYRENE/BUTADIENE RUBBER (SBR)

TABLE 2.7
SBR PRODUCTION CAPACITIES IN 1989

Country	Thousands of tons/year	
North America	**1190**	
United States		1040
Latin America	310	
Brazil		250
Western Europe	1320	
Italy		320
West Germany		270
United Kingdom		240
France		220
Netherlands		140
Spain		85
Belgium		45
Australia	35	
India	40	
Japan	700	
Taiwan	120	
Miscellaneous Africa and Asia	225	
CPEC[1]	1760	
World	**5700**	

1. Centrally-planned economy countries.

Chapter 3

POLYBUTADIENE

The industrial development of polybutadiene dates from the early 1960s with the advent of stereoregular elastomers with a majority cis-1,4 configuration.

World production is now close to one million tons (excluding centrally-planned economy countries).

Many processes have been developed, characterized by solution polymerization (emulsion polymerization processes having virtually disappeared) and which mainly differ in the type of catalyst system used.

3.1 GENERAL INFORMATION
ON POLYBUTADIENES

Butadiene can theoretically polymerize in four different ways (Fig. 3.1).

Polymerization can occur by 1,4 addition: the remaining double bond may be either cis or trans.

If polymerization occurs by end-to-end 1,2 addition, two stereoisomers are possible, depending on the arrangement of the asymmetrical carbon atoms. If they all have the same d or l configuration, the polymer is said to be isotactic. If they have an alternate d and l configuration, the polymer is syndiotactic.

If polymerization occurs by 1,4 and 1,2 addition at random, the polymer is said to be atactic. It cannot crystallize, as opposed to a polymer whose configuration approaches 100% 1,4-cis or trans.

3.2 POLYMERIZATION CHEMISTRY

The polymerization of butadiene is distinguished by high exothermicity ($\Delta H = -17.6$ kcal/mol).

3.2.1 Monomer Purity

The butadiene specifications are the same as those required for the solution copolymerization of SBR.

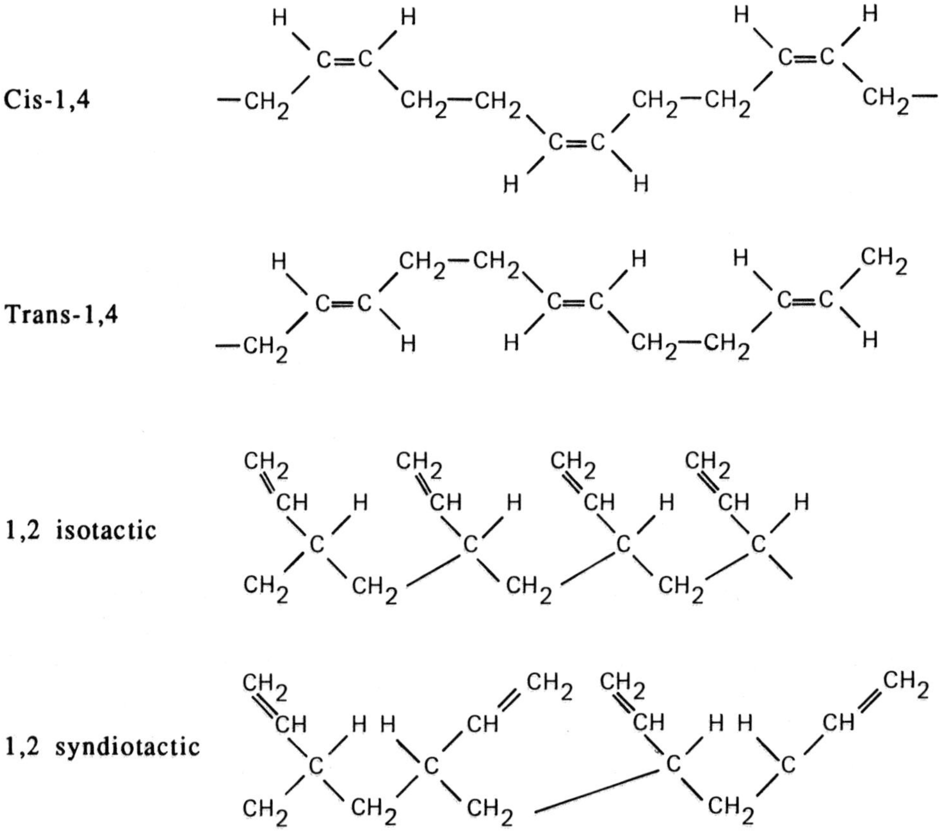

Fig. 3.1 Stereoregular polybutadiene structures.

J.P. ARLIE

3.2.2 Catalysts

Four main catalyst systems are found in industrial units, which are derivatives of one or two metals. These catalysts are distinguished by their solu-
bility in the reaction medium. Catalyst systems of the Ziegler/Natta type are
used mainly to produce cis-1,4 polybutadiene. These catalysts are based on
the salt of a transition metal combined with a reducing organometallic compound.

Associated with the main catalyst systems is a cis-1,4 molecular configuration, as shown in Table 3.1.

TABLE 3.1

CATALYST SYSTEMS AND CONFIGURATION OF POLYBUTADIENE CHAINS

Catalyst system	Cis-1,4 percentage of polybutadiene
$CoCl_2$-R_2AlCl	97 to 98
Cobalt acetylacetonate/$(C_2H_5)_2AlCl$	96 to 98
AlR_3-$AlCl_3$-$CoCl_2$	94 to 95
Nickel naphthenate/$(C_2H_5)_3Al$/$BF_3(C_2H_5)O$	97 to 98
$TiI_4(C_2H_5)_3Al$/$(C_3H_7)_2O$	91 to 94
Butyl lithium	40 to 60

3.2.3 Molecular Weight

The weight average molecular weight $\overline{M}_w$ of the polybutadienes range between $3 . 10^5$ and $5 . 10^5$. The distribution curve depends on the catalyst system, as shown by Fig. 3.2. Catalyst systems based on lithium derivatives yield
polymers with a narrower distribution curve. The advantage of such a system
over the Ziegler systems is to compensate for the lower cis-1,4 microstructure.
They present the drawback of difficult processing of the gum.

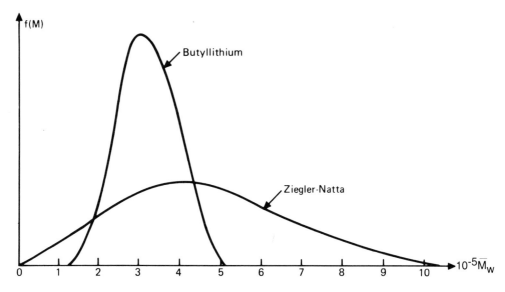

Fig. 3.2 Molecular weight distribution curve of polybutadiene as a function of catalyst system used.

3.3 POLYMERIZATION PROCESSES

Apart from the emulsion process that is still used today by one company in the United States *(Texas US Chemical Company)*, current techniques operate in solution (Fig. 3.3).

Whatever the catalyst system, all these processes include the following four steps:

(a) Polymerization.
(b) Solvent and unconverted monomer recovery.
(c) Polymer recovery.
(d) Elastomer drying and packaging.

The process licensors, classed by the main metal in the catalyst, are:

Ziegler/cobalt:
Shell, Goodrich, Gulf, Montecatini (Italy), *Hüls* (West Germany).

Ziegler/titanium:
Phillips (USA), *Bayer* (West Germany), *Goodyear, Esso, Goodrich, Gulf, Polymer Corporation.*

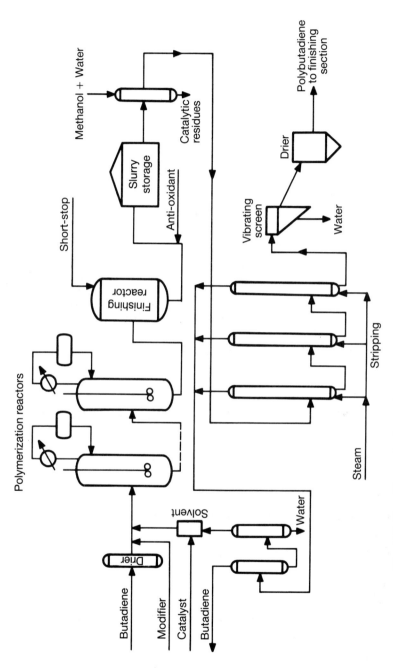

Fig. 3.3 Butadiene solution polymerization process. Catalyst system: Ziegler/cobalt.

Nickel:
Bridgestone Tire Company (Japan), *Japan Synthetic Rubber Company* (Japan).

Butyl lithium:
Firestone Tire and Rubber Company.

3.3.1 Polymerization

Polymerization is continuous. The monomer stabilizer, usually tertbutyl-catechol, is removed by washing the butadiene with caustic soda solution. This is then removed by washing with water. The butadiene is then dried on molecular sieves to remove any trace of moisture. The same treatment is applied to the solvent, which serves as a diluent for the catalyst.

A technical comparison of the butadiene polymerization step for the different industrial processes is shown in Table 3.2.

TABLE 3.2

TECHNICAL COMPARISON OF BUTADIENE POLYMERIZATION PROCESSES

Polymerization conditions	Catalyst system			
	Ziegler/ titanium	Ziegler/ cobalt	Nickel	Lithium
Temperature (°C)	0.0	5.0	40.0	50.0
Pressure (bar)	3.5	0.3	3.5	3.5
Solvent	Toluene + butene-1	Benzene	Toluene	Hexane
Conversion per pass (%)	60.0	80.0	87.0	97.0
Time (h)	2.0	5.0	2.0	4.0
Reactor volume (m^3)	15.0	23.0	15.0	15.0
Number of reactors/line	9.0	4.0	4.0	6.0
Number of lines for production of 50,000 t/year	2.0	4.0	2.0	2.0
Solvent/monomer ratio in feed	13.0	4.3	6.0	4.0
Percentage of polymer in slurry	4.5	15.0	12.0	19.0

The catalyst is mixed with the monomer and solvent before being fed to the reactor. In this way, the catalyst is totally and uniformly introduced into the feed and recycle butadiene. The amount introduced can be adjusted in accordance with the conversion or molecular weight of the desired polymer.

The polymerization reaction takes place in a series of reactors at least 15 m^3 in volume. These reactors are made of vitrified steel. Despite its relatively low thermal conductivity, vitrified steel proves to be better than many materials, because the vitrified layer is thin. The resulting surface is extremely smooth, which hinders or prevents any polymer from adhering. Heat transfer resistance thus remains constant throughout the reaction, and this is important for the polymerization of butadiene. These reactors are normally equipped with a jacket to remove the heat of reaction. Rapid stirring is provided by a turbine stirrer.

The reactor dimensions are selected to have a high height to diameter ratio, close to 4, to ensure a larger cooling surface.

In some processes, the heat of reaction is removed by vaporization, followed by condensation of the solvent in an external condenser, after which the solvent is returned to the reaction medium.

The latest reactors have been specially designed to remove the heat of reaction of highly viscous media.

Depending on these processes, after a few hours have elapsed, the polymer concentration may exceed 15%. The reaction is then stopped in an auxiliary reactor by the addition of a suitable short-stop, such as a fatty acid. An antioxidant (phenyl β naphthylamine) is added to the polymer solution or slurry.

In processes using cobalt-based catalysts, the catalyst concentration is such that it demands the removal of catalyst residues to avoid subsequent oxidation and ageing. This is done by hot-stirring the slurry with a dilute solution of methanol in water.

The rubber solution is then sent to a series of mixing tanks for homogenization.

3.3.2 Monomer, Solvent and Polymer Recovery

This recovery takes place in the steam-stripping section which comprises one or more strippers, in which the unconverted butadiene, C_4 impurities in the feed and solvent are extracted at the top by steam injection at 100°C, while the elastomer coagulates in small particles. Steam stripping explains the high steam expenditure in solution processes. This quantity of steam depends on the percentage of solid rubber leaving the reactors.

The mixture of solvent, monomer and water is distilled. The water leaves at the column bottom. The hydrocarbons, solvent and monomer are recovered at the top, dried on an alumina bed, and recycled to the polymerization section.

3.3.3 Drying and Packaging

After filtration on a vibrating screen, the polymer gum is extruded to remove any trace of liquid and then dried with hot air. The extremely pure elastomer obtained is packed in approximately 35 kg bales.

3.4 ECONOMIC DATA

Table 3.3 shows the economic data concerning processes using cobalt- or lithium-based catalysts. It gives average figures valid for Western Europe.

Besides the economic comparison, processes using butyl lithium display the following differences in technical characteristics from those using cobalt:

(a) Higher conversion for a smaller amount of catalyst, hence no problem or removal of catalytic residues.
(b) The polymer contains no gel for 100% conversion, even for polymers with a weight average molecular weight as high as 500,000.
(c) The drawback stems from the greater difficulty of processing the polymer, which is due to the absence of low molecular weights (see curve in Fig. 3.2).

3.   POLYBUTADIENE

TABLE 3.3
SOLUTION PROCESSES FOR POLYBUTADIENE,
ECONOMIC DATA

	Catalyst system	
	Cobalt	Lithium
Capacity (t/year)	90,000	90,000
Battery limits investments $(10^6$ FF 1989)	570	450
Consumption per ton of polybutadiene		
• **Raw material:**		
Butadiene (t)	1.01	1.01
• **Solvents:**		
Hexane (kg)	–	30
Benzene (kg)	40	–
• **Chemicals and catalysts:**		
Butene-1 (kg)	3	
Methanol (kg)	1	
Stabilizer (kg)	10	10
Catalysts (kg)	10	
N-butyl lithium (kg)		0.5
Caustic soda (kg)		0.3
Palmitic acid (kg)		5
• **Utilities:**		
Steam (t)	10	8.5
Cooling water (m^3)	210	170
Process water (m^3)	7	2
Electricity (kWh)	670	800
Labor (operators per shift)	10	10

3.5 USES AND PRODUCERS

The main use of polybutadiene is in the tire industry. It is also used as an additive to polystyrene to form impact grade polystyrene. The market in the United States in 1989 was broken down as follows:

Use	Percentage	
Tires	77	
Automobiles		43
Utility vehicles		27
Impact polystyrene	19	
Miscellaneous	4	

Polybutadiene is used in tires to increase tread wear resistance. It is employed in a blend with SBR or natural rubber. About 1 kg is consumed per tire for automobiles and 3.3 kg for utility vehicles.

The growth of radial-casing tires should favor polybutadiene, which can be used in tire side walls because of its capacity to eliminate internal heating.

Polybutadiene production capacities in 1989 are given in Table 3.4.

3.6 OTHER POLYBUTADIENES

Other elastomers (specialty elastomers) and polybutadiene resins include the following:

(a) Vinyl polybutadienes with a proportion of 1,2 units ranging between 20 and 80%. They are produced by solution polymerization using butyl lithium and a chelatant ether such as tetrahydrofuran in heptane. These elastomers, which are distinguished by high moisture resistance, good thermal stability and a low dielectric constant, behave like thermosetting plastics after vulcanization. Their main applications are in electricity and electronics.

(b) Non-stereoregular low molecular weight liquid polybutadiene resins are mainly produced in the United States. These polymers, whose terminal groups are OH or COOH groups, are highly reactive, especially with diisocyanates, yielding foams or specialty rubbers.

TABLE 3.4

PRODUCTION CAPACITIES OF CIS-1,4 POLYBUTADIENE IN 1989

Country	Thousands of tons/year	
North America	510	
United States		400
Latin America	55	
Brazil		55
Western Europe	455	
France		190
West Germany		70
Italy		50
United Kingdom		60
Spain		20
Australia	40	
India	30	
Japan	260	
Centrally-planned economy countries	385	
World	**1800**	

Chapter **4**

POLYISOPRENE

This elastomer is a substitute for natural rubber. It has been produced industrially since 1963 by stereospecific polymerization of isoprene.

World production is 800,000 tons annually, including more than 80% for the centrally-planned economy countries. This relatively small production in comparison with SBR can be explained by the high cost of isoprene, which currently prevents the production of polyisoprene at cost lower than the market price of natural rubber. Only the CPEC countries, which consume less natural rubber for strategic reasons (15% of world rubber consumption) have developed and accelerated the production of polyisoprene.

Isoprene polymerization processes operate in solution and are continuous. The catalyst systems used are either Ziegler systems or lithium derivatives. They yield a stereoregular structure of polyisoprene containing more than 90% of cis-1,4 units.

4.1 GENERAL INFORMATION ON ISOPRENES

Like butadiene, isoprene can be polymerized in four different ways (Fig. 4.1).

The main configuration desired in cis-1,4 is that of natural rubber. The trans-1,4 configuration corresponds to a plastic, which is similar to the natural product gutta percha.

J.P. ARLIE

Cis-1,4

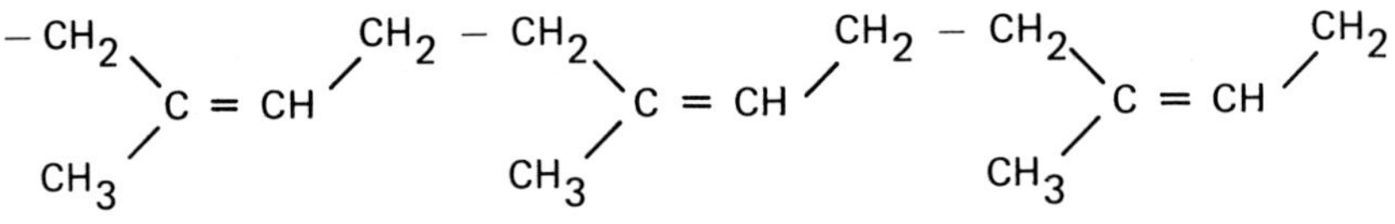

$- CH_2$
$C = CH$
CH_3
$CH_2 - CH_2$
$C = CH$
CH_3
$CH_2 - CH_2$
$C = CH$
CH_3
CH_2

Trans-1,4

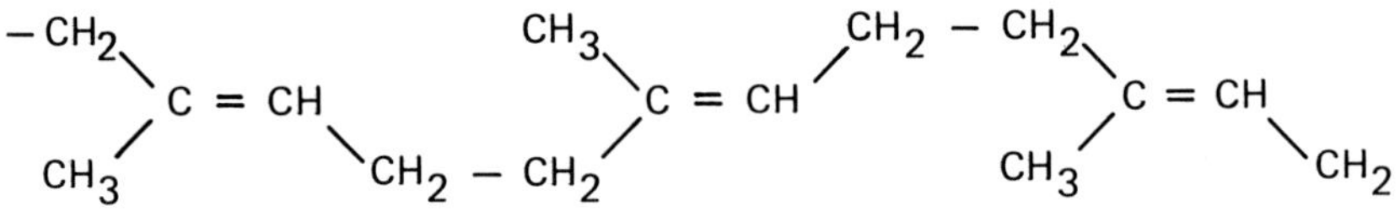

$- CH_2$
$C = CH$
CH_3
$CH_2 - CH_2$
CH_3
$C = CH$
$CH_2 - CH_2$
$C = CH$
CH_3
CH_2

Trans-1,2

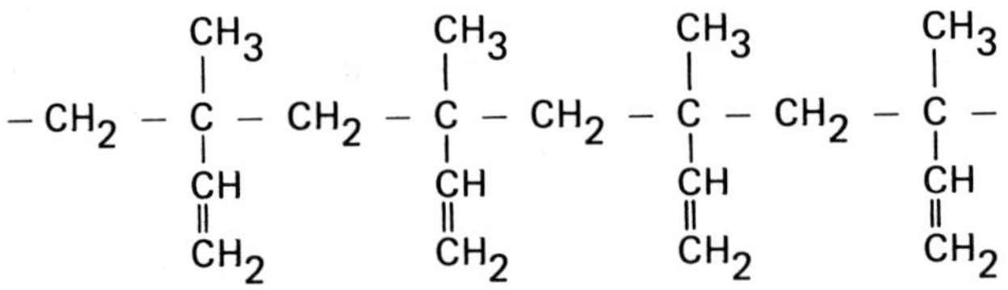

$$- CH_2 - \underset{\underset{CH_2}{\overset{||}{CH}}}{\overset{CH_3}{\underset{|}{C}}} - CH_2 - \underset{\underset{CH_2}{\overset{||}{CH}}}{\overset{CH_3}{\underset{|}{C}}} - CH_2 - \underset{\underset{CH_2}{\overset{||}{CH}}}{\overset{CH_3}{\underset{|}{C}}} - CH_2 - \underset{\underset{CH_2}{\overset{||}{CH}}}{\overset{CH_3}{\underset{|}{C}}} -$$

Trans-3,4

$$- CH - CH_2 - CH - CH_2 - CH - CH_2 -$$
$$\underset{CH_3 \quad CH_2}{C} \quad \underset{CH_3 \quad CH_2}{C} \quad \underset{CH_3 \quad CH_2}{C}$$

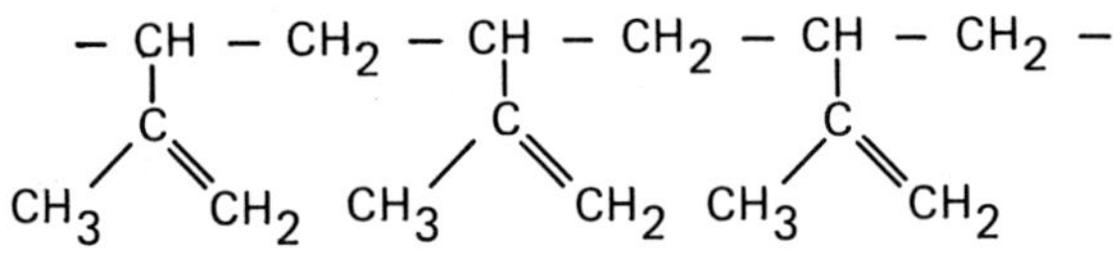

Fig. 4.1 Microstructures of polyisoprenes.

4.2 POLYMERIZATION CHEMISTRY

The heat of polymerization of isoprene is 17.9 kcal/mol.

4.2.1 Monomer Purity

Monomer purity varies in accordance with the polymerization catalyst. The specifications required are shown in Table 4.1.

TABLE 4.1

ISOPRENE SPECIFICATIONS FOR POLYMERIZATION

	Catalyst	
	Ziegler	Butyl lithium
Isoprene (% weight)	99.5 minimum	99 minimum
Olefins (% weight)	–	1 maximum
Cyclopentadiene (ppm)	50 maximum	1 maximum
Piperylene (ppm)	150 maximum	100 maximum
Acetylenics (ppm)	50 maximum	20 maximum
Inhibitor (ppm)	200 to 300 maximum	–

4.2.2 Catalysts

Two types are encountered, both soluble in the reaction medium: Ziegler/ Natta catalyst systems and lithium derivatives.

Table 4.2 shows the microstructure of the polyisoprenes obtained in accordance with the catalyst system employed. A comparison is made with the micro-structure found in "smoked sheet No.1", a characteristic sample of natural rubber.

The Ziegler catalyst is prepared by the combination of triisobutyl aluminum with titanium tetrachloride to obtain an aluminum to titanium ratio between 0.9 and 1.1, which yields the maximum of cis-1,4 units and high molecular weights for the polymer.

The addition of diphenyl ether or trialkyl aluminum increases conversion per pass of isoprene.

TABLE 4.2

COMPARISON OF THE MICROSTRUCTURE OF SYNTHETIC POLYISOPRENES
AND NATURAL RUBBER

	Synthetic rubbers		Natural rubber
	Lithium catalyst	Ziegler catalyst	Smoked sheet No.1
Cis-1,4	92.5	96.0	97.0
Trans-1,4	1.5	0.5	1.0
1,2	–	–	–
3,4	6.0	3.5	2.0
Total	100.0	100.0	100.0

4.2.3 Molecular Weight

The molecular weight ranges between 100,000 and 150,000. The polydispersity index is lower for polymers obtained by processes using lithium-based catalyst systems.

4.3 POLYMERIZATION PROCESSES

Two processes are encountered, depending on the catalyst system. The main process licensors are:

(a) Titanium/aluminium (Ziegler): *Goodyear, Goodrich, SNAM Progetti.*
(b) Butyl lithium: *Shell.*

Units capacities are close to 25,000 t/year.

These continuous solution polymerization processes are very similar to those found in the polymerization of butadiene (Chapter 3). They also include four steps: polymerization, polymer recovery, recovery and recycle of solvent and unreacted monomer, and polymer drying and packing (Fig. 4.2).

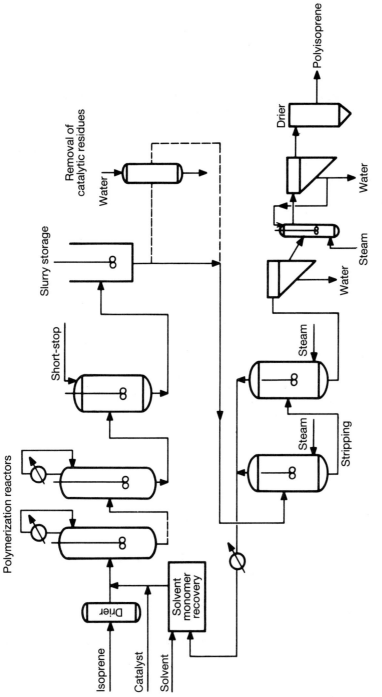

Fig. 4.2 Solution polymerization of isoprene, Ziegler catalyst.

The operating conditions generally encountered are shown in Table 4.3. The main differences concern the method used to remove the heat generated by the reaction and the need to eliminate the catalytic residues in the Ziegler processes. In technologies based on butyl lithium catalysis, the heat is removed by the vaporization of isopentane and part of the isoprene, and recondensation in the reactor by means of an external condenser. In processes based on Ziegler catalysis, part of the heat of reaction from the first reactor is decreased by cooling with the aid of the feed (solvent + monomer), and the excess heat is removed through the reactor jacket.

TABLE 4.3

TECHNICAL COMPARISON OF ISOPRENE POLYMERIZATION PROCESSES

Polymerization conditions	Catalyst system	
	Ziegler	Butyl lithium
Temperature (°C)	50	55
Pressure (bar)	1	1.5
Solvent	Hexane	Isopentane
Conversion per pass	80	75
Time (h)	2	2
Reactor volume (m^3)	15	15
Number of reactors per line	4	4
Number of lines for production of 50,000 t/year	2	2
Solvent/monomer ratio in feed	4.3	4
Percentage of polymer in slurry	15	15

4.4 ECONOMIC DATA

Table 4.4 gives the average figures for polymerization processes using butyl lithium and Ziegler catalysts.

POLYISOPRENE - ECONOMIC DATA

Capacity (t/year)	40,000
Battery limits investments (10^6 FF 1979)	480
Consumption per ton of polyisoprene:	
Raw material: isoprene (t)	1.0
Catalysts and chemicals (F)	210
Utilities:	
HP steam (t)	7
Cooling water (m^3)	10
Electricity (kWh)	720
Labor (operators per shift)	10

4.5 USES AND PRODUCERS

The properties of cis-1,4 polyisoprene are very similar to those of natural rubber. These elastomers are characterized by high tensile and yield strength. They withstand low temperatures, but not as well as polybutadiene, and they are easily blended with various ingredients. Their internal heating is low due to good resilience, but they age poorly and display a tendency to crack. They display poor resistance to oils, ozone, adverse weather and abrasion.

The main uses of polyisoprene are approximately the following in the United States:

(a) Tires   60%
(b) Mechanical objects   16%
(c) Shoe soles   7%
(d) Sports articles and toys   4%
(e) Miscellaneous   13%

4. POLYISOPRENE

Production capacities in 1989 for different countries are shown in Table 4.5.

TABLE 4.5

PRODUCTION CAPACITIES OF POLYISOPRENE IN 1989

Country	Thousands of tons
United States	65
Netherlands	40
South Africa	45
Japan	70
CPEC	1045
World	**1265**

Chapter **5**

ETHYLENE/PROPYLENE
RUBBERS

Ethylene/propylene rubbers are among the latest synthetic elastomers, with development work dating from 1961/1962 and the first industrial production by the *Enjay Chemical Company (Exxon)* in 1962. Basic research on the synthesis and properties of these rubbers was conducted by G. Natta and *Montecatini*, who demonstrated that, by copolymerization and with the aid of Ziegler-type catalysts, ethylene and propylene yielded an amorphous product with interesting elastomer properties.

World production was about 500,000 t in 1989. The processes are either solution or suspension polymerization processes.

5.1 GENERAL INFORMATION ON
ETHYLENE/PROPYLENE RUBBERS

Ethylene/propylene rubbers are generally represented by the acronym EPDM, in which E represents ethylene, P propylene, D the diene required in a small proportion for vulcanization, and M the methylene units (CH_2), which form the hydrocarbon skeleton of the copolymer. If the copolymer contains no diene, it can be denoted by EPM. The acronym EPR (ethylene/propylene rubber) is the most common, and includes both copolymers and terpolymers.

5.1.1 Polymerization Chemistry

Ethylene and propylene are copolymerized in an organic solvent in the liquid phase. The heat of polymerization depends on the composition of the co-polymer. The data for polyethylene and polypropylene are as follows:

$$nC_2H_4 \rightarrow (C_2H_4)_n \qquad \qquad \Delta H°_{298} = -25.88 \text{ kcal}$$

$$nC_3H_6 \rightarrow (C_3H_6)_n \qquad \qquad \Delta H°_{298} = -24.89 \text{ kcal}$$

The weight average molecular weight ranges between 10^5 and $2 . 10^5$.

5.1.2 Molecular Configuration

The structure of alternate ethylene/propylene copolymers is written:

$$\underset{}{(CH_2 - CH_2 - \underset{\underset{CH_3}{|}}{CH} - CH_2)_n}$$

This structure, which is identical to that of cis-1,4 polyisoprene, is never found in industrial rubbers. Even for an equimolecular ethylene/propylene composition, the configuration is not alternate: the polymer contains short chains (or blocks) of polyethylene and polypropylene, which are dispersed at random, in the midst of longer segments of statistical ethylene/propylene copolymers.

The macromolecules are not perfectly linear and are made up, in various proportions, of branches of short and long chains. These structural variations are influenced by the conditions of polymerization and the ethylene/propylene composition, as well as diene for EPDM terpolymers.

The composition of industrial products is usually given in weight per cent. Although the elastomer properties of ethylene/propylene copolymers are obtained over a wide range of compositions, the products actually vary in composition from 50/50 to 75/25 ethylene/propylene. Those with the highest propylene content are easier to process, whereas, in the opposite case, the physical properties of the gum are improved. The composition also influences the crystal-linity, which must be as close to zero as possible. Towards the upper ethylene limit, the copolymer becomes crystalline, making it insoluble. A higher ethylene content favors the lowering of the glass transition temperature and accordingly improves the low temperature behavior of the copolymer.

5.1.3 Choice of the Termonomer

Ethylene/propylene copolymers have the drawback of not containing double bonds, and therefore cannot be vulcanized by sulfur. The vulcanization of these products hence requires costly peroxides, which are unusual in the rubber processing industries. Unsaturations are introduced into the chain by adding a termonomer which contains two double bonds with different reactivities, with the less active not entering the polymer chain. Three aliphatic dienes are used for the purpose, 1,4-hexadiene, dicyclopentadiene and ethylidene norbornene, and the latter is probably the most widely used.

The structure of an EPDM containing ethylidene norbornene is the follow-ing:

$$\left(CH_2 - CH_2 - \underset{\underset{CH_3}{|}}{C} \quad CH_2 \quad \right)_n$$

$$\underset{CH_3}{\overset{\parallel}{CH}}$$

The quantity of diene incorporated in the EPDM considerably affects the vulcanization rate, and 4 to 5% by weight is usually sufficient. Some grades designed for very high vulcanization rates may contain up to 10%.

5.1.4 Monomer Purity

Monomer purity must be examined closely. The moisture content in par-ticular must be close to one part per million. Typical specifications of ethylene and propylene are given in Table 5.1.

TABLE 5.1
SPECIFICATION OF ETHYLENE AND PROPYLENE FOR EPR RUBBER

Ethylene:

Ethylene (weight %)	>99
Saturated hydrocarbons (weight %)	<1
Propylene and heavy products (ppm by volume)	40
CO_2 (ppm by volume)	5
Acetylene (ppm by weight)	1
Sulfur (ppm by weight)	1
H_2O (ppm by weight)	<1
CO (ppm by volume)	2
Oxygen (ppm by weight)	1
Hydrogen (ppm by weight)	1
Chlorine (ppm by weight)	<5

Propylene:

Propylene (weight %)	>99
Saturated hydrocarbons (weight %)	<1
CO_2 (ppm by volume)	<5
CO (ppm by volume)	<2
Methylacetylene (ppm by weight)	<1
Allene (ppm by weight)	<5
Oxygen (ppm by volume)	<5
Hydrogen (ppm by volume)	<20
Sulfur (ppm by volume)	<1
H_2O (ppm by volume)	<5
Cl_2 (ppm by volume)	<5
Other unsaturates (ppm by volume)	<20

5.1.5 Catalysts

The catalyst systems are derived from Ziegler catalysts. Several catalyst combinations have been reported, and the most typical use a catalyst that is a chloride of a transition metal, vanadium, and a co-catalyst, alkylaluminum:

(a) $VOCl_3$ $(C_2H_5)_3 AlCl_3$
(b) $VOCl_3$ $(C_2H_5)_2 AlCl$
(c) $VOCl_3$ $(C_4H_9)_2 AlCl$

These catalysts are liquids transportable in tanks. They must be handled in the total absence of air. They are normally diluted in a hydrocarbon such as hexane.

5.2 POLYMERIZATION PROCESSES

Two types of process are used depending on the reaction medium.

Processes operating in a solvent medium were developed by *Esso Research and Engineering Company (Exxon)*. Suspension processes were developed by *Montecatini Edison*.

The operations are continuous and highly automated.

5.2.1 Solution Processes

One of the features that distinguishes solution polymerization processes from other monomers is operation in the liquid phase. The solvent is hexane.

Another original feature of these techniques is the combination of the catalyst VOC_3 and co-catalyst $[(C_2H_5)_3 AlCl_3]$ with the three monomers in a vigorously stirred mixer just before introduction into the reactor. By performing this operation in a very short interval, in a small volume, at low temperature ($-40°C$), the catalyst system and monomers are perfectly distributed.

The polymerization unit for a capacity of 30,000 t/year has two reactors that can be used separately. Each reactor is equipped with a high-capacity stirrer.

The solution of catalysts and monomers in hexane is introduced into the jacketed reactor in which the temperature is kept between 30 and 40°C under 15 bar during the reaction. Hexane serves to reduce the viscosity of the medium. Hot acidic water is then introduced to facilitate the extraction of the catalyst residues, an operation which is indispensable to reduce the ash content of the polymer in order to avoid subsequent accelerated ageing of the gum.

Conversion per pass is about 50% for ethylene, 10% for propylene and 80% for the diene.

The demineralized solution is sent to two evaporators, where unreacted monomers are recovered by hot water and steam injection. The polymer is dispersed in water in the form of small particles. The hexane, ethylene, propylene and diene are redistilled, dried and recycled.

The solution containing about 10% polymer is dried in four successive phases:

(a) Removal of a large part of the water (50%) on dewatering screen.
(b) Concentration to 10% water in an mechanical screw press.
(c) Heating of the elastomer under pressure in an expander until the moisture content is 1%.
(d) Final drying in a expander-dryer.

At the discharge end, the rubber grains fall into a hopper which feeds a press, in which they are weighed and pressed into bales.

5.2.2 Suspension Process

The suspension process also have five steps: polymerization (liquid propylene is used as diluent), removal of catalytic residues using a solvent (toluene), steam stripping to recover the solvent and unreacted monomers, recycle of the monomers and solvent, and drying of the polymer (Fig. 5.1).

The main differences between this process and solution processes concern the removal of the heat of polymerization and the quantity of solvent/diluent to be recycled.

The heat of reaction is removed by evaporation of the monomers from the reaction phase, followed by compression, condensation and recycle of the condensate to the reactor.

Due to the rapid increase in viscosity of the reaction medium with the increase in polymer concentration, the polymer cannot exceed 10% at the end of the reaction in solution processes. In suspension processes, the solid polymer has little effect on the viscosity of the reaction medium, so that the polymer concentration can be up to 30%, achieving a gain in the amount of diluent to be recycled.

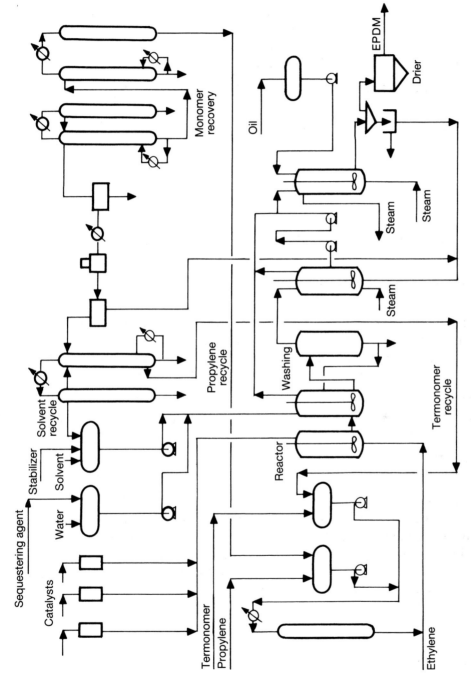

Fig. 5.1 Production of ethylene/propylene rubber by emulsion process.

5.3 ECONOMIC DATA

Tables 5.2 and 5.3 give information concerning suspension and solution copolymerization processes. The battery limits investments are given for a capacity of 45,000 t/year, which is the average size observed.

TABLE 5.2

ETHYLENE/PROPYLENE RUBBER

ECONOMIC DATA ON SUSPENSION PROCESS

Capacity (t/year)	45,000
Battery limits investments: 10^6 FF 1989	280
Consumption per ton of EPDM rubber:	
• Raw materials:	
Ethylene (t)	0.615
Propylene (t)	0.350
Ethylidene norbornene (t)	0.070
• Catalysts and chemicals (FF)	480
• Utilities:	
Steam (t)	4
Cooling water (m^3)	330
Process water (m^3)	4
Electricity (kWh)	1000
Labor (operators per shift)	8

TABLE 5.3

ETHYLENE/PROPYLENE RUBBER

ECONOMIC DATA ON SOLUTION PROCESS ACCORDING TO THE TERMONOMER USED

Capacity (t/year)	45,000	45,000
Battery limits investments $(10^6$ FF 1989)	370	250
Consumption per ton of EPDM rubber:		
• Raw materials:		
Ethylene (t)	0.68	0.65
Propylene (t)	0.28	0.34
Ethylidene norbornene (t)	0.07	–
1,4-hexadiene (t)	–	0.04
• Catalysts and chemicals:		
Vanadium derivative (kg)	1.30	2.60
Aluminum derivative (kg)	9.70	10.60
N-hexane (kg)	20.00	5.00
Aniline (kg)	–	5.50
Polypropylene glycol (kg)	23.00	–
Hydrogen (kg)	–	0.08
Caustic soda (kg)	14.30	0.01
• Utilities:		
Steam (t)	22	2
Cooling water (m^3)	880	300
Process water (m^3)	19	0.1
Electricity (kWh)	1050	960
Labor (operators per shift)	6	7

5.4 PROPERTIES

Terpolymers can be vulcanized by sulfur, but, since the double bond of the diene is not in the main chain, but only in a side chain, the oxidizing agents can attack this double bond without touching the main chain. Vulcanized terpolymers accordingly exhibit excellent resistance to air, ozone and corrosive chemicals. They display good abrasion resistance as well as excellent dielectric properties. They show poor resistance to combustion, but good resistance to heat (up to 100°C).

Another important characteristic from the economic standpoint is that, in the preparation of mixtures for finished products, these elastomers accept high percentages of filler (carbon black and silica) and oil.

5.5   USES AND PRODUCERS

The main applications concern six major industry sectors:

(a) Automobile manufacture (this sector accounts for about 70% of all uses): external body seals, water circulation pipes, etc. Tire applications are very limited.
(b) Building and public works: all types of seal.
(c) Piping industry: steam, compressed air, garden hoses.
(d) Electrical switchgear: insulation of electrical cables (protective sheath or stuffing).
(e) Home appliances: washing machine parts, for example.
(d) Lubricating oils: viscosity index (VI) improvers.

The production capacities of ethylene/propylene rubber in 1989 are given in Table 5.4.

TABLE 5.4

PRODUCTION CAPACITY OF ETHYLENE/PROPYLENE RUBBER IN 1989
(IN THOUSANDS OF TONS PER YEAR)

USA	270
Japan	135
USSR	135
Italy	60
Netherlands	55
West Germany	41
France	65
World total	**811**

Chapter 6

BUTYL RUBBER

Butyl rubber has been produced industrially since 1942 by *Esso Standard Oil (Exxon)* in the United States.

Standard Oil 's basic research demonstrated that, by using Friedel/Craft type catalysts, and by operating at low temperature, it was possible to polymerize a mixture of isobutene (97 to 98.5% depending on grade) and a conjugated diene (3 to 1.5%). The unsaturation resulting from the low percentage of diene introduced into the polymer chain easily allows subsequent vulcanization of the copolymer obtained.

World production in 1989 was approximately 500,000 t/year.

6.1 GENERAL INFORMATION ON BUTYL RUBBER

Isobutene and isoprene are copolymerized in an organic solvent at low temperature (–100°C) by a cationic polymerization process.

The molecular configuration of butyl rubber can be represented as follows:

$$\left[(CH_2 - \underset{\underset{CH_3}{|}}{\overset{\overset{CH_3}{|}}{C}})_n - (CH_2 - \underset{}{\overset{\overset{CH_3}{|}}{C}} = CH - CH_2)_m \right]_x$$

where the ratio n/n+m ranges between 0.97 and 0.995. The isoprene molecules, which are distributed at random in the polymer chain, mainly have the trans-1,4 configuration.

The average molecular weights range between 300,000 and 450,000, and the polydispersity index measured by the gel permeation chromatography method is about 3. The molecular weights depend both on the polymerization temperature and the respective concentrations of the reactants. Thus the logarithm of the molecular weight varies directly and linearly with the inverse of the absolute polymerization temperature. Since isoprene reacts much slower than isobutene, its incorporation in the chain reduces the molecular weight. The impurities in the feed, such as butenes, cause chain transfer and termination reactions. Hence the monomers must be very pure (minimum 99%). The specifications of isoprene are given in Chapter 4 in Table 4.1.

The specifications of the isobutene recovered are those obtained after extraction of the naphtha steam-cracking C_4 cut using sulfuric acid. Butane and butene impurities are about 1%.

6.2 POLYMERIZATION PROCESSES

Existing processes are based on the technology developed by *Exxon* (Fig. 6.1). These are continuous solution processes, at low temperature. The production of butyl rubber displays the following specific requirements:

(a) The need to keep the composition of the feed constant in terms of isoprene and isobutene, because isoprene reacts more slowly than isobutene.
(b) The need to keep the temperature of the reactor at –100°C, requiring special alloys, and good insulation to prevent losses during cooling: this is achieved by means of liquid ethylene.
(c) The very high rate of reaction, with complete copolymerization in less than 1 s.

The starting solution contains about 25% by weight of isobutene and 0.5 to 1.5% isoprene diluted in methyl chloride. The reaction takes place at –100°C in the presence of 0.2% by weight of $AlCl_3$ as catalyst. Instantaneous polymerization occurs in a cylindrical reactor surrounded by a tubular heat exchanger which serves to cool the solution and remove the heat. Annual

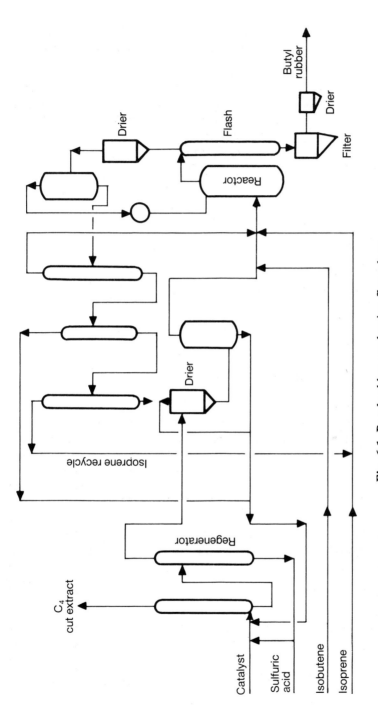

Fig. 6.1 Butyl rubber production flow chart.

production of 40,000 tons requires six reactors (unit volume 15 m^3). Due to the deposition of the polymer on the internal surface of the reactor, which disturbs the heat transfers, it is indispensable to provide an auxiliary reactor for two reactors, allowing for longer production time without interruption for cleaning. To avoid subsequent deposits in the flash tanks during solvent separation, vigorous stirring is necessary, combined with the addition of zinc stearate to minimize the caking of the polymer particles. The polymer in a 3% solution in water is stabilized by the addition of phenyl naphthylamine. The water is then removed by vacuum filtration and then by a tunnel drier. Since butyl rubber displays a stronger tendency to low-temperature flow than other elastomers, it is pressed in the form of bales covered by a polyethylene film to prevent the adhesion of the polymer to any other surface during storage and transport.

6.3 ECONOMIC DATA

The economic data are given in Table 6.1. Investments concern a capacity of 40,000 t/year representative of the average in the United States and Western Europe.

TABLE 6.1

ECONOMIC DATA FOR BUTYL RUBBER

Capacity (t/year)	40,000
Battery limits investments: 10^6 FF 1989	460
Consumption per ton of butyl rubber:	
• Raw materials:	
Pure isobutene (t)	1.010
Isoprene (t)	0.025
• Catalyst and chemicals (FF)	540
• Utilities:	
Steam (t)	11
Cooling water (m^3)	975
Process water (m^3)	10
Electricity (kWh)	850
Labor (operators per shift)	11

6.4   PROPERTIES

The low percentage of isoprene which makes vulcanization possible (not as easy as for natural rubber) helps to obtain a practically saturated product which is unaffected by external causes of deterioration: heat, light, oxygen, ozone, acids. In addition to its properties, the side cluttering of the polymer chain by the CH_3 groups of isobutene imparts excellent impermeability to this elastomer (ten times greater than that of natural rubber and other elastomers). Butyl rubber also has excellent electrical properties, especially its insulation resistance, as well as dielectric properties. Suitably plasticized, it remains elastic at very low temperature to around −60°C.

6.5   USES AND PRODUCERS

Depending on the isobutene/isoprene ratio, the varieties of rubber obtained display more or less accentuated resistance to ageing and ozone. Several grades are available, corresponding to the various uses, which differ in the isobutene/ isoprene ratio and the Mooney viscosity (parameter associated with the average molecular weight).

The main applications are:

(a) Inner tubes for pneumatic tires, miscellaneous mechanical goods, gas pipes (propane, butane), coated fabrics.
(b) Articles exposed to weather and sunlight.
(c) Insulation for cables.
(d) Articles for high-temperature applications (belts, pipes, gaskets, radiator hoses).
(e) Articles resistant to chemicals (pipes for the chemical industry) and fatty materials (pipes for the food industry).

In the unvulcanized state, butyl rubber is used to make glues and adhesives, special mastics, and to plasticize waxes and polyethylene.

One limitation for the general uses of butyl rubber stems from the difficulties of blending with other elastomers, due to the wide differences in the vulcanization reaction.

Production capacities recorded in 1989 are shown in Table 6.2.

6. BUTYL RUBBER

TABLE 6.2

PRODUCTION CAPACITY IN 1989 OF BUTYL RUBBER
(IN THOUSANDS OF TONS PER YEAR)

USA	219
Belgium	85
Japan	82
United Kingdom	60
France	48
Canada	120
CPEC	120
World total	**734**

Chapter **7**

POLYCHLOROPRENE RUBBER

Polychloroprene has been commercialized since 1932 by *Du Pont de Nemours* under the brand name Neoprene. It is a special-purpose rubber whose world consumption represents about 5% of the market for synthetic elastomers. The monomer is chloroprene or 2-chloro 1,3-butadiene.

7.1 CHLOROPRENE PREPARATION AND CHEMISTRY

In the earliest industrial units, chloroprene was obtained from acetylene, but, since 1960, with the development of the petrochemical industry, new plants are based on chloroprene obtained by the indirect chlorination of butadiene.

7.1.1 Chloroprene from Acetylene

The dimerization of acetylene to monovinyl acetylene (MVA) is the first step in the polymerization of acetylene. This reaction takes place in the presence of cuprous chloride:

$$2\ HC \equiv CH \xrightarrow{\ CuCl\ } CH \equiv C - C = CH_2 \xrightarrow[HC \equiv CH]{\ CuCl\ } CH_2 = CH - C \equiv C - CH - CH_2 + \text{isomers}$$

The conversion per pass of acetylene is low. This gives a good yield of MVA, but the acetylene must be recycled and the MVA separated. The reactor is vitrified, the pressure is 5 bar, and the temperatures ranges between 25 and 80°C. MVA is separated by preferential absorption followed by vaporization. The MVA yield is better than 90%.

MVA is converted to chloroprene by the addition of hydrochloric acid in the presence of cuprous chloride. The reaction yields chloroallene, which is rapidly rearranged into 2-chloro 1,3-butadiene:

$$CH_2 = CH - C \equiv CH + HCl \xrightarrow{CuCl} CH_2Cl - CH = C = CH_2$$

$$CH_2Cl - CH = C = CH_2 \xrightarrow{} CH_2 = CH - CCl = CH_2$$

$$CH_2 = CH - CCl = CH_2 + HCl \xrightarrow{CuCl} CH_2Cl - CH = CCl - CH_3$$

The total yield of this two-step process is about 90%.

7.1.2 Chloroprene from Butadiene (Fig. 7.1)

The synthesis takes place in three steps.

7.1.2.1 Chlorination of Butadiene

$$CH_2 = CH - CH = CH_2 + Cl_2 \longrightarrow \begin{cases} ClCH_2 - CH = CH - CH_2Cl \\ \\ ClCH_2 - CHCl - CH = CH_2 \end{cases}$$

This takes place in the vapor phase at around 290 to 330°C. Conversion per pass is less than 25%, and the dichlorobutene yield is about 90%. The products obtained at the reactor exit are chlorobutadiene-1 and -2, dichlorobutene-3,4, dichlorobutene-1,4, and higher boiling point products including trichlorobutenes, tetrachlorobutanes, telomers and tars. These products are separated under reduced pressure to minimize the decomposition of the most volatile components.

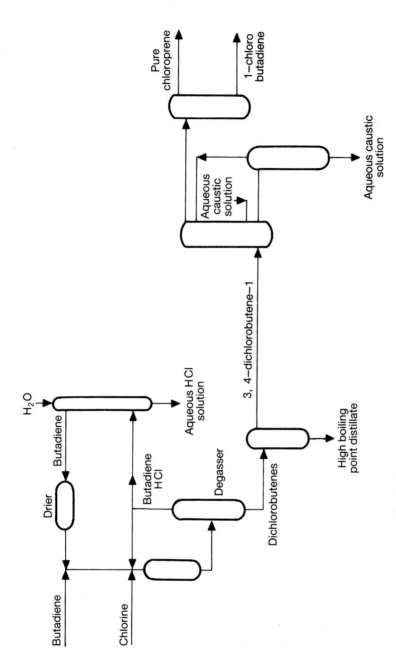

Fig. 7.1 Preparation of chloroprene from butadiene and chlorine.

7.1.2.2 Isomerization of Dichlorobutenes

If the exclusive use of the dichlorobutenes is chloroprene, dichloro-1,4 butene-2 must be isomerized to dichloro-3,4 butene-1. At 100°C, in the presence of a catalyst (CuCl), the mixture of dichlorobutenes is balanced at 21% of 3,4, 7% of cis-1,4, and 72% of dichlorobutene-1,4. Since the vapor phase contains these isomers, dichlorobutene-3,4 can be isolated from this phase, due to its lower boiling point. The 1,4 dichlorobutenes are recycled to the liquid phase containing the catalyst.

7.1.2.3 Dehydrohalogenation of Dichloro-3,4 Butene-1

This reaction takes place in the presence of dilute caustic soda in a stirred reactor at about 100°C. After the caustic solution is removed, the effluent is sent to the distillation section to recover unreacted dichlorobutene and to recycle it to the reactor. Final purification after drying takes place under reduced pressure to remove any traces of impurities and, in particular, to decrease the amount of chlorobutadiene which may reach 3%, an acceptable level for polymerization. The yield of this step is generally higher than 95%.

The purified monomer can be polymerized directly, or stored at low temperature (2°C) in the presence of a polymerization inhibitor (tert butyl catechol). It is stored at this temperature under nitrogen atmosphere.

7.2 CHLOROPRENE POLYMERIZATION

This operation, which uses the free radical emulsion method, is similar to styrene/butadiene copolymerization (Chapter 2). The processes are batch for small capacities, or continuous for units of 40,000 t/year or more.

Chloroprene purity is not a major problem, for the simple reason that the polymerization rate is extremely high, so that inhibition reactions do not occur.

7.2.1 Chloroprene Polymerization Chemistry

Like butadiene and isoprene, chloroprene polymerizes with a large release of heat (17 kcal/mol). Polymerization can take place unpredictably due to various influences: traces of impurities playing the role of initiator, temperature, and light.

Chloroprene also displays a high affinity for oxygen. In the presence of oxygen which serves as a polymerization initiator, a highly cross-linked polymer is formed, called popcorn due to its appearance.

7.2.1.1 Chemical Agents Employed

A number of anionic soaps of fatty acids (colophane salt), alkyl sulfates or alkylaryl sulfonates are generally used as emulsifiers. The choice of the soap depends on the final product desired.

Potassium or ammonium persulfate is the most widely used promoter.

7.2.1.2 Molecular Weight

The polymerization of chloroprene yields a polymer with a very high molecular weight, a highly branched and strongly gelled structure, making its subsequent processing very difficult. A number of modifiers is used to reduce the molecular weights. Two families of polymers are thus distinguished:

(a) Sulfur-modified polychloroprenes.
(b) Mercaptan-modified polychloroprenes.

Mercaptans act as polymerization regulators due to their chain transfer function. Unlike sulfur-modified polymers, mercaptan-modified polymers do not have bonds liable to be broken by mechanical work or thermal activation.

The number average molecular weights are higher than 50,000. The polydispersity defined by $\overline{M_w}/\overline{M_n}$ ranges between 2 and 5. These data relate to the "SOL" polychloroprenes that as soluble in specific solvents, contrary to "GEL" polychloroprenes which are insoluble in these solvents. The latter are elastomers consisting of an assembly of three-dimensional zones, with weak mutual interactions and hence capable of sliding over each other, explaining the specific and reduced uses of GEL polymers in comparison with the uses of SOL polymers.

7.2.1.3 Molecular Configuration

The molecular configuration is essentially determined by the polymerization temperature and by the types of modifier used. Chloro-2 butadiene tends to yield a preferential trans-1,4 configuration, without the need for stereospecific catalysts. Polychloroprene is the only industrial synthetic elastomer obtained by the free radical method to display a stereospecific character.

The high rate of trans-1,4 units results in a pronounced tendency to crystallization and the excellent resulting mechanical properties.

7.2.2 Manufacturing Process

The polymerization of chloroprene in the stage of industrial manufacture includes the same steps as the emulsion polymerization of styrene/butadiene (SBR).

The process licensors manufacturing polymerization chloroprene are *Du Pont* (USA) and *British Petroleum/Distillers* (United Kingdom).

Polymerization operations are conducted in autoclave reactors placed in series, with a unit volume up to 20 m³. These reactors are equipped with a stirrer and jacket to favor heat exchanges with the reactants. The coolant is either water or a cold brine, depending on the polymerization temperature.

After weighing, the raw material is fed to the reactors in the presence of de-ionized and de-aerated water and solutions of emulsifiers.

A typical formula for copolymerization in the presence of sulfur is the following, expressed in parts by weight:

Chloroprene	100	
Resinic acid	4	Dissolved in
Sulfur	0.6	the monomer
Water	150	
Caustic soda	0.8	
Sodium methylene naphthenate	0.7	
Potassium persulfate	0.2 to 1	

The sodium salt of the resin is formed in situ. Polymerization takes place at $40 \pm 0.5°C$, at atmospheric pressure, by the addition of potassium persulfate at a steady rate, so that the conversion rate of the monomer follows a linear program as a function of time. As soon as conversion reaches 90%, a level determined by measuring the latex density, the reaction is stopped by the addition of 2.5 parts of tetraethyl thiurame disulfide $[(C_2H_5)_2 NCS]_2 S_2$. This product destroys the excess initiator and combines with all the free radicals present to yield stable macromolecules. In polymers obtained by sulfur modification, each latex particle is a cross-linked macromolecule which subsequently undergoes peptization, an operation in which the polysulfide bridges are broken to eliminate chain cross-linking. The unconverted monomer is then stripped with steam, and then recovered by condensation.

The elastomer is extracted from the latex by coagulation in a few seconds at temperatures between –10 and –15°C. This delicate operation is one of the specific features of chloroprene polymerization. It takes place at low temperature to minimize dehydrochlorination. Coagulation is continuous on a rotating drum cooled by internal coolant circulation. After washing and drying, the film is transformed into a strand 3 cm in diameter, and then cut and packed in 25 kg bags.

7.3 ECONOMIC DATA

Tables 7.1 and 7.2 give the economic data on plants of 35,000 t/year, which represent the average production capacity reported in the literature.

TABLE 7.1

ECONOMIC DATA RELATED TO CHLOROPRENE

	Acetylene process	Butadiene process
Capacity (t/year)	35,000	35,000
Battery limits investment: 10^6 FF 1989	115	80
Consumption per ton of chloroprene:		
• Raw materials:		
Acetylene (t)	0.68	
Hydrochloric acid (t)	0.49	
Butadiene (t)		0.71
Chlorine (t)		0.90
Caustic soda (t)		0.66
• Catalysts and chemicals (FF)	800	850
• Utilities:		
Steam (t)	4	2.7
Cooling water (m^3)	218	175
Electricity (kWh)	100	70
Labor (operators per shift)	5	5

TABLE 7.2
ECONOMIC DATA RELATED TO POLYCHLOROPRENE

Capacity (t/year)	35,000
Battery limits investment: 10^6 FF 1989	140
Consumption per ton of polychloroprene:	
• Raw materials: chloroprene (g)	0.96
• Catalysts and chemicals:	
Mercaptan (kg)	3
Sulfur (kg)	2
Resinic acid (kg)	30
Initiator and other products (kg)	
• Utilities:	
Steam (t)	5.5
Cooling water (m^3)	300
Electricity (kWh)	5
Refrigeration (10^3 frigories)	0.4
Labor (operators per shift)	14

(1)   Combined production of sulfur- and mercaptan-modified grades.

7.4   USES AND PRODUCERS

The mechanical properties of polychloroprene are very similar to those of natural rubber, but the existence of the chlorine atom decreases the sensitivity of the double bond to oxidizing agents.

The main features of this rubber are:

(a) Resistance to ozone.
(b) Resistance to oxidation by air, and hence to ageing.
(c) Self-extinguishing thanks to the presence of chlorine.
(d) Excellent resistance to oils and greases.
(e) Good low-temperature behavior and excellent adhesive properties.

The main applications of the solid rubber are the following:

(a) Wires and cables.
(b) Manufacture of standard sections (automotive, building).

(c) Flexible hoses (petroleum industry).
(d) Conveyor and transmission belts.
(e) Coated fabrics.

Applications in the form of latex concern gloves, mechanical parts, weather balloons, the impregnation of paper and cardboard, self-extinguishing foams for furniture, coating of fabrics and underneath of carpets.

This rubber is also useful for the preparation of many adhesives adaptable to a number of applications: the building, shoe, furniture and automobile industries.

The main market sectors are listed in Table 7.3.

TABLE 7.3

MARKET FOR POLYCHLOROPRENE RUBBER IN WESTERN EUROPE IN 1987

Use	% consumption
Automotive (belts, hoses)	32
Consumer products (conveyor belts, seals, wet suits)	22
Adhesives	20
Coating of wires and cables	14
Miscellaneous	12

Production capacities are given in Table 7.4.

TABLE 7.4

PRODUCTION CAPACITIES OF POLYCHLOROPRENE RUBBER IN 1989
(IN THOUSANDS OF TONS PER YEAR)

USA	163
Japan	85
West Germany	60
France	40
United Kingdom	35
CPEC	150
World total	**563**

Chapter **8**

NITRILE RUBBER

Nitrile rubber is a copolymer of butadiene and acrylonitrile. Produced in the 1940s under the brand names Perbunan and Buna N, it is now denoted by the acronym NBR. The production of nitrile rubber represents about 3% of world production of synthetic elastomers.

The polymerization processes are similar to those described for emulsion SBR. The acrylonitrile content of the polymer ranges between 18 and 50%, with a value of about 32% by weight.

The polymerization processes are batch processes.

8.1 BUTADIENE/ACRYLONITRILE COPOLYMERIZATION CHEMISTRY

Nitrile rubber is produced by the following reaction:

$$\left[CH_2 = CH - CH = CH_2 \right]_x \quad + \quad \left[CH_2 = CH - CN \right]_y \longrightarrow$$

$$\left\{ CH_2 - CH = CH - CH_2 \right\}_x \left\{ CH_2 - CHCN \right\}_y$$

In addition to this main reaction, butadiene can react partly at 1,2, yielding the following copolymer:

$$\longrightarrow \left\{ CH_2 \cdot CH = CH - CH_2 \right\}_{x-z} \left\{ CH_2 - CH - CN \right\}_y \left\{ CH_2 - CH \right\}_z$$
$$\begin{array}{c} | \\ CH \\ \| \\ CH_2 \end{array}$$

These vinyl groups can in turn polymerize to form side chains or branches.

8.1.1 Influence of Acrylonitrile Content

In addition to the possibility of vulcanization, the butadiene group imparts suppleness and flexibility at low temperature. The acrylonitrile group imparts resistance to hydrocarbons and impermeability to gases.

As the acrylonitrile content increases, resistance to hydrocarbons, impermeability to gases and thermal resistance also increase. By contrast, the resilience, low-temperature flexibility and viscosity of the solutions decrease. **About 80% of all industrially-manufactured nitrile rubbers have an acrylonitrile content between 24 and 35%.**

8.1.2 Monomer Purity

The specifications of butadiene are those required for the synthesis of SBR. Those of acrylonitrile are shown in Table 8.1.

TABLE 8.1

SPECIFICATIONS OF ACRYLONITRILE

Aldehydes	≤20 ppm
Hydrocyanic acid	<5 ppm
Iron	≤0.03 ppm
Acetonitrile	<500 ppm
Peroxides	< 0.1 ppm
Copper	≤0.03 ppm
Divinyl acetylene	Undetectable
Methyl vinyl ketone	Undetectable
Cyanobutadiene	Undetectable
Non-volatile matter	≤60 ppm

8.1.3 Molecular Configuration

For a product containing 28% acrylonitrile, the percentage of 1,2 vinyl units of butadiene is 10.5% against 89.5% in the 1,4 configuration. This proportion drops slightly as the percentage of acrylonitrile increases, but it remains independent of the polymerization temperature.

8.1.4 Molecular Weight

Molecular weight is a parameter that influences the processing characteristics. A high molecular weight, which is reflected by a high Mooney viscosity, leads to a product that can accept large amounts of plasticizers. Contrary to natural rubber, nitrile rubber does not depolymerize during mastication. Depending on the qualities and grades desired, the Mooney viscosity, measured at 100°C, may vary from 30 to 100.

8.2 NITRILE RUBBER SYNTHESIS PROCESSES

As for SBR, nitrile rubber was manufactured for many years by the hot emulsion polymerization method. This was supplanted in recent years by the cold emulsion polymerization process.

Contrary to the technology for obtaining SBR, **most of the nitrile rubber synthesis processes are batch processes**, and this offers an advantage for the manufacture of several grades of elastomer for special purposes.

Since the basic technique of *I.G. Farben* has been in the public domain for several years, each manufacturer can propose a license, whether *Polysar*, *Goodrich*, *Goodyear*, *Uniroyal*, *Bayer* or *PCUK*.

A typical polymerization recipe is shown below, expressed in parts by weight:

Butadiene	75
Acrylonitrile	25
Water	180
Soap	4.5
Stearic acid	0.6
Dodecyl mercaptan	0.5
Potassium chloride	0.3
Sodium pyrophosphate	0.1
Ferric sulfate	0.02
20% hydrogen peroxide	0.35

The volume of the autoclave reactors for polymerization can be as high as 20 m³, and the temperatures range between 20 and 30°C. Reaction time varies from 10 to 24 h. Polymerization is generally stopped at 80% conversion, hence slightly higher than that obtained for SBR. These higher conversions are possible because of the lower viscosity of the reaction medium. The endstop is either hydroquinone or sodium sulfide. The latex is stabilized by phenyl β-naphthylamine, and then coagulated, by aluminum sulfate for example. The subsequent separation and drying steps are identical to those described for SBR. The products are delivered in 25 kg bales.

8.3 ECONOMIC DATA

Since some production plants for SBR rubber are used to produce nitrile rubber, the economic data concern a 35,000 t/year unit capable of producing either SBR or NBR (Table 8.2).

TABLE 8.2

ECONOMIC DATA FOR NITRILE RUBBER

Capacity (t/year)	35,000
Battery limits investments: 10^6 FF 1989	240
Consumption per ton of nitrile rubber:	
• Raw materials[1]:	
Butadiene (t)	0.35
Acrylonitrile (t)	0.65
• Chemicals and promoter:	
Miscellaneous chemicals (kg)	25
Sodium chloride (kg)	200
Soap (kg)	70
Sulfuric acid (kg)	24
• Utilities:	
Steam (t)	3.5
Cooling water (m³)	180
Process water (m³)	460
Electricity (kWh)	5
Labor (operators per shift)	17

(1) Average figures.

8.4 USES AND PRODUCERS

Nitrile rubber is distinguished by excellent resistance to oils, gasoline and greases, as well as many solvents and chemicals. In general, nitrile rubber is used whenever good hydrocarbon resistance is important.

The main use sectors are the following:

(a) Automotive and aeronautical construction, which account for 65% of commercial outlets: all parts expected to be in permanent or accidental contact with fuels, oils or greases, such as O-rings, radiator hoses, and pressed cork gaskets.
(b) Hydrocarbon handling and transport: hoses for tanker loading and unloading, flexible tanks, valve and piping gaskets.
(c) Rollers and cylinder lining in the textile, paper and printing industries.
(d) Safety shoe soles.
(e) Adhesives for cementing rubber, and for cementing plasticized polyvinyl chloride on various supports.

In addition to these main applications, the presence of the polar groups of nitrile rubber imparts excellent compatibility with a number of plastics such as polyvinyl chloride. Table 8.3 gives he production capacities in 1989.

TABLE 8.3

PRODUCTION CAPACITIES OF NITRILE RUBBER IN 1989
(IN THOUSANDS OF TONS PER YEAR)

USA	113
Canada	25
France	41
West Germany	48
United Kingdom	16
Italy	45
Netherlands	50
Japan	75
Latin America	16
CPEC	97
World total	**552**

Chapter **9**

THERMOPLASTIC ELASTOMERS

In the past ten years, elastomers that can be processed by techniques specific to thermoplastics have assumed greater commercial importance. These products can be used in areas that demand a number of rubbery properties combined with the economy resulting from rapid processing, which characterizes the plastics industry and, accordingly, avoid the vulcanization step necessary for elastomers.

The original properties of thermoplastics stem from the presence of rigid blocks or segments, that behave as thermally labile physical bonds which soften and flow under the action of shear forces or heat, and then resume their initial structure after the force has been released or by cooling.

The **first type** of these block copolymers consists of styrene/butadiene/styrene or styrene/isoprene/styrene copolymers marketed by *Shell* in 1965. The rigid segment is represented by the vitreous phase of the agglomerated blocks of polystyrene, which are incompatible with the diene blocks, creating a two-phase structure in the polymer, one continuous and preponderant, with the properties of an elastomer, while the second, which is the discontinuous minority phase, displays the properties of a thermoplastic.

The **second type** owes its reinforcing character to the hydrogen bonds between the polymer chains. This applies to thermoplastic polyurethanes, of which the hydrogen bonds result from the urethane functional groups.

In the third type, the physical strength stems from the crystallizable polyester blocks. These form crystalline domains dispersed in an amorphous phase. Examples are ether copolyesters.

A number of significant examples of elastomers are listed in Table 9.1

World production, which was close to 400,000 tons in 1989, is expected to grow at a faster rate than the other elastomers in the years to come.

TABLE 9.1

EXAMPLES OF THERMOPLASTIC ELASTOMERS

Thermoplastic elastomer	Hard block	Soft block
Styrene block copolymer (SBC)	Polystyrene	Polybutadiene Polyisoprene
Polyurethane block copolymer (TPU)	Polyurethane	Polyester Polyether
Polyester block copolymer (COPE)	Polyester	Polyether
Polyolefin blends (TPO)	Polyethylene Polypropylene	Ethylene/ propylene rubber
Polyether block amide (PBA)	Polyamide	Polyether
Styrene rubber/thermoplastic resin blends (IPN)	Polytetra- fluoroethylene	Styrene, ethylene/ butene, styrene

9.1 GENERAL INFORMATION ON THERMOPLASTIC ELASTOMERS

Thermoplastic elastomers are polymers which can be stretched to at least twice their initial length, and which return to their initial dimension as soon as the stretching force is released, without any vulcanization.

The vitreous or crystalline blocks, which behave like structural cross-linking points, soften under the action of heat like thermoplastics. After returning to ambient temperature, the polystyrene blocks recombine to produce a rigid article. The elastomer phase imparts its rubber properties to the unit, while the vitreous phase confers the tensile strength.

The chief advantage of these products is that they can be processed by equipment used to handle thermoplastics. However, this advantage is offset by the higher cost of these elastomers in comparison with other synthetic elastomers.

Thermoplastic elastomers can be obtained by chemical reaction (copolymerization) or by physical blending of an elastomer with a crystalline compound. Only the former method is dealt with here, because the latter concerns blends of EPDM rubber (Chapter 5) with polyethylene or polypropylene (TPO). The final product contains about 60 to 80% EPDM elastomer.

Thermoplastic elastomers may also contain various agents such as glass fibers, lubricants, etc.

9.2 STYRENE/DIENE BLOCK COPOLYMERS

These copolymers account for the largest share of the market for thermoplastic elastomers. Marketed by *Shell* in 1965, they are made up of a two-phase system in which the vitreous polystyrene domains are linked by elastomer chains incompatible with polystyrene, which consist of polybutadiene (SBS), polyisoprene (SIS), or ethylene/butene copolymer (SEBS). The molecular configuration may be linear or branched (three or four branches).

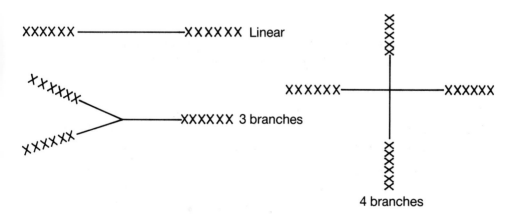

Linear block copolymers are marketed by *Shell* under the trade name Kraton or Cariflex.

Branched block copolymers are marketed by *Phillips Petroleum* under the name Solprene or Europrene *(Anic)*.

9.2.1 Copolymerization Techniques

The synthesis of block copolymers based on styrene is an application of the homogeneous phase anionic polymerization initiated by lithium-based compounds. This technique helps to approximate accurately the desired composition and molecular weight.

In the free radical polymerization reactions, chain growth is interrupted by mutual reaction of two chains, by hydrogen transfer from one chain to another, by transfer from the monomer to the solvent, or by the fixation of impurities on the free radical ends.

Anionic polymerization does not present most of these drawbacks. Transfer reactions in particular are less frequent. However, the most important feature of this type of polymerization is the chain termination mode. By selecting a suitable anionic initiator, such as butyl lithium, and by taking care to use extremely pure reactants, it is possible to inhibit the termination reactions and to preserve the reactive ends of the macromolecules. These polymers are qualified as "living". The polymerization, which is stopped by the depletion of the monomer, resumes if a new quantity is added. It can also resume if another monomer is added, carefully selected for its redox potential. Thus, with monomers of varied electron affinities, it is possible to produce polymers with several blocks.

Control of the chain termination reaction helps to obtain products distinguished by a very narrow molecular weight distribution.

Bifunctional initiators, such as compounds of lithium and sodium, can be used to prepare A-B-A linear block copolymers:

$$2\,CH_2 = CH - CH = CH_2 + Li - R - Li \longrightarrow$$

Diene   Initiator

$$Li - CH_2 - CH = CH - CH_2 - R - CH_2 - CH = CH - CH_2 - Li$$

"Living chain"

9. THERMOPLASTIC ELASTOMERS

$$(n + m)CH_2 = CH - CH = CH_2 + Li - R - Li \longrightarrow$$

$$Li(CH_2 - CH = CH - CH_2)_n R(CH_2 - CH = CH - CH_2)_m Li$$

Polydiene

$$(x + y) \, CH = CH_2 + \text{Polydiene} \longrightarrow Li(CH - CH_2)_x \, (\,\text{Block}\,) \, (CH - CH_2)_y \, Li$$
$$\text{polydiene}$$

Styrene   Polystyrene block   Polystyrene block

"Living" block copolymer

The initiators, which are generally soluble in ether or polar solvents, give rise to polydiene blocks with a low 1,4 microstructure content, which has the effect of diminishing the mechanical properties at low temperature by the increase in the glass transition temperature.

The difficulty in the preparation of styrene/diene block polymers is that styrene and dienes have comparable electro-affinities. However, since the reaction of the styrene ion with isoprene is faster than the reaction of the isoprenyl ion with styrene, the styrene polymerizes first.

To increase the content of cis-1,4 units in the polydiene block, organolithium initiators and non-polar solvents are selected, which raises a number of problems, because the organolithium initiators are insoluble in non-polar media, and the slow initiations always cause fluctuations in the copolymer composition. The following processes are available:

(a) Butyl lithium + styrene, followed by diene addition, and then styrene addition with a solvent such as tetrahydrofuran to accelerate initiation.
(b) Butyl lithium + styrene, then addition of half of the required quantity of diene, and coupling by a bifunctional electrophilic compound: coupling can be obtained in polar medium, and examples of coupling are shown in Table 9.2.
(c) Butyl lithium + styrene, then addition of a blend of diene and styrene: the addition of styrene/diene/styrene three-block copolymer is feasible in these conditions, but a transition zone exists between the two blocks, containing diene and styrene blocks.

TABLE 9.2

SYNTHESIS OF STYRENE/DIENE BLOCK COPOLYMERS BY THE COUPLING PROCESS

Diene	Promoter	Solvent	Coupling agent
Isoprene	Sec butyl lithium	Cyclohexane	Dibromo-1,2 ethane
Isoprene	Sec butyl lithium	Cyclohexane +isopentane	Dibromo-1,2 ethane
Isoprene	Sec butyl lithium	Cyclohexane	Ethyl acetate
Isoprene	Sec butyl lithium	Benzene	Dibromo-1,3 propane
Butadiene	Sec butyl lithium	Cyclohexane	Vinyl acetate
Butadiene	N-butyl lithium	Cyclohexane	Dimethyldichlorosilane
Butadiene	N-butyl lithium	Cyclohexane	Epoxidized polybutadiene
Butadiene	N-butyl lithium	Cyclohexane	Divinyl benzene
Butadiene	Sec butyl lithium	Cyclohexane	Diethyl adipate

9.2.2 Copolymerization Initiators

Several organometallic compounds are capable of initiating copolymerization without a termination reaction. The most important are the organolithium compounds soluble in hydrocarbons and ethers.

Every promoter anion is capable of initiating a polymer chain, so that the quantity of initiator determines the chain length or the molecular weight of the polymer ($\overline{M}n$):

$$\overline{M}_n = \frac{\text{Weight of monomer in grams}}{\text{Number of molecules of initiator}}$$

The most widely used promoters are organolithium compounds, and their relative reactivities are as follows:

(a) **For dienes:**
   sec butyl lithium > isopropyl lithium > n-butyl lithium and ethyl lithium.
(b) **For styrene:**
   sec butyl lithium > isopropyl lithium > n-butyl lithium and ethyl lithium.
(c) **Solvent effect:**
   toluene > benzene > n-hexane > cyclohexane.

9.2.3 Coupling Agents and Operating Conditions

The reaction time and temperature may vary in accordance with the coupling agent employed. The pressure factor is less important: the operating pressures are sufficient to keep the monomers in the liquid phase.

A number of examples of operating conditions are listed in Table 9.3. The quantity of coupling agent required by the reaction is generally equal to the stoichiometric quantity.

The desired molecular weight is about 100,000 for the copolymer.

TABLE 9.3
COUPLING AGENTS AND OPERATING CONDITIONS

Coupling agent	End product	Temperature	Time
Ethyl acetate	Styrene/isoprene/styrene	80°C	1 h
Epoxidized polybutadiene	Styrene/butadiene (with grafting)	50°C	Instantaneous
Vinyl acetate	Styrene/butadiene/styrene	60°C	1 h
Ethyl acetate	Styrene/butadiene/styrene	25 to 60°C	2 h

9.2.4 Copolymerization Processes

The most important thermoplastic elastomers are the styrene/butadiene/styrene block copolymers in which the polybutadiene block has a composition with 55% 1,4-polybutadiene, and 35% 1,2-butadiene. The process uses the coupling method.

The manufacturing process is a semi-continuous process, because the polymerization reaction is a batch process, whereas solvent and polymer recovery are continuous.

Batch polymerization offers the advantage of better control of the chemical reactions and greater flexibility for the production or different product grades.

The main drawback is the need for very large reactors with which the removal of the heat of reaction is more difficult.

To limit reactor size and to produce different grades of copolymers simultaneously, a 50,000 t/year unit was selected. It has five reaction lines in parallel (Fig. 9.1). The process includes three separate steps:

(a) Styrene polymerization.
(b) Diene polymerization.
(c) Coupling of "living polymer" chains.

9.2.4.1 Styrene Polymerization

The purified solvent (cyclohexane) and monomer are pumped to the first reactor, which is kept under nitrogen atmosphere. The temperature is about 35°C. The solution of promoter (1% by weight of butyl lithium in cyclohexane) is rapidly fed into the reaction mixture. The time required to transfer the initiator must be as short as possible, with maximum homogenization of the mixture. These two parameters condition the quality of the product, with respect to its narrower molecular weight distribution.

Any impurity equivalent to a proton donor, such as water, destroys an equivalent of butyl lithium, which leads to a higher molecular weight product.

As soon as the initiator enters into contact with the styrene, a styryl lithium is formed and polymerization begins. It is conducted adiabatically. The temperature in the reactor rises to 55°C due to the exothermicity of the reaction (160 kcal . kg).

After one hour, the styrene is completely converted to polystyrene, and the solution containing 10% of "living" polymer in the solvent is rapidly transferred to the reactor, which contains the diene in solution in cyclohexane. There is no problem of heat transfer in this step.

9.2.4.2 Diene Polymerization

Before transferring the solution of "living" polystyrene, the solvent and diene (butadiene or isoprene), previously purified, are introduced into a second reactor placed under nitrogen atmosphere. The temperature of the mixture is 55°C.

If no polar compound is present in the solution, the diene polymerization reaction mainly yields the 1,4 polybutadiene configuration. This reaction takes 2 h, and the temperature is kept close to between 50 and 60°C. The heat of polymerization of butadiene (340 kcal/kg) or isoprene (260 kcal/kg) is removed by a combination of top-distillate condenser, reactor jacket and in-

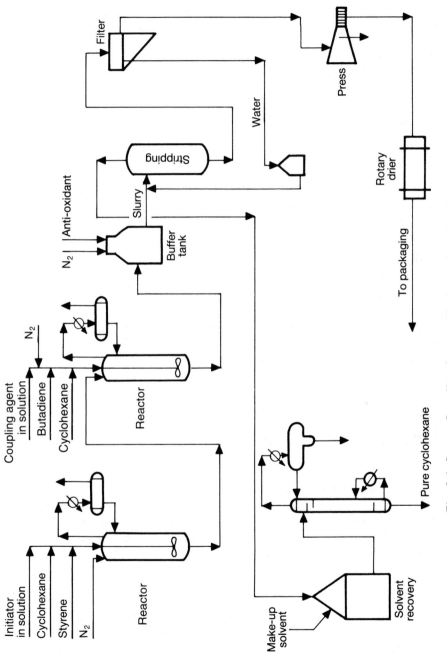

Fig. 9.1 Styrene/butadiene/styrene block copolymerization process.

ternal cooling coil (cooled by water). As the reaction advances, the reaction rate declines, and the heat of polymerization can then be removed by the jacket and the cooling coil. At the end of the reaction, the reactor contains a block polymer of the type:

Polystyrene/polydiene/Li

9.2.4.3 Coupling of "Living Polymer" Chains

A stoichiometric quantity of solution of coupling agent (5% by weight in cyclohexane) is introduced into the reactor, and the mixture is then heated to 80°C for 1 h. This reaction leads either to the linear copolymer with three blocks by using a bifunctional coupling agent, or the branched copolymer by using a multi-functional coupling agent.

The block copolymer is transferred to a buffer tank which serves as a link between the batch section (polymerization) and the continuous section (solvent and polymer recovery).

9.2.4.4 Solvent and Polymer Recovery

The slurry containing 15% by weight of polymer, mixed with a solution of anti-oxidant, is sent to the coagulation section. Most of the solvent is removed by stripping. The polymer leaving the coagulation section consists of 20% (by weight) of solid in suspension in an aqueous phase containing the residues of initiator and coupling agent. After filtration and extrusion, the polymer, which still contains 5% water, is sent to the drying section (rotary drier operating at 70°C). The product is then sent to the storage and packing area.

The cyclohexane recovery section includes a drying column and a fractionation column. The cyclohexane is thus rectified on a bed of molecular sieves to remove any trace of moisture before recycle.

9.2.5 Economic Data

The economic data given in Table 9.4 correspond to production capacities of 40,000 t/year, consisting of four lines of reactors of 10,000 t/year each.

TABLE 9.4

ECONOMIC DATA ON STYRENE/BUTADIENE STAR BLOCK COPOLYMER

Capacity (t/year)	40,000
Battery limits investments: 10^6 FF in 1989	240
Consumption per ton of thermoplastic rubber:	
• Raw materials:	
Styrene (t)	0.295
Butadiene (t)	0.695
• Catalysts and chemicals:	
N-butyl lithium (kg)	1.5
Cyclohexane (kg)	20.0
Miscellaneous chemicals (kg)	20.0
Tetrahydrofuran (kg)	1.0
• Utilities:	
Steam (t)	8
Cooling water (m^3)	45
Electricity (kWh)	600
Labor (operators per shift)	9

9.3 POLYURETHANE-BASED THERMOPLASTIC ELASTOMERS

Polyurethanes are obtained by the reaction of a polyester diisocyanate or polyol polyether with a diol.

Polyurethane-based thermoplastic elastomers consist of rigid blocks (low molecular weight copolymer consisting of a diol and an isocyanate) and flexible blocks (polyester or polyol polyether). At ambient temperature, the rigid blocks form hydrogen bonds which cross-link the flexible blocks, giving rise to the rubbery behavior observed. At higher temperature, the hydrogen bonds are broken, and the material assumes the behavior of a thermoplastic.

9.3.1 Basic Raw Materials

Polyurethane-based block copolymers are prepared by mixing a diisocyanate with a diol and a polyol in stoichiometric proportions. The diol generally used is 1,4-butanediol and the diisocyanate is methane diphenyl -p-p' diiso-cyanate (MDI).

In preparing the polyester, the polyol selected is poly(oxytetramethylene adipate)glycol, with a molecular weight between 800 and 1100. The mixture of diol and polyol must be adjusted to obtain a molecular weight of about 500.

The polyester is prepared by reaction of 1,4-butanediol with adipic acid.

In preparing the polyether, the polyol selected is polyol selected is poly(oxytetramethylene ether)glycol, with a molecular weight between 1000 and 2000. The average molecular weight of the mixture of diol and polyol is about 500.

Polyurethane-based thermoplastic elastomers are generally prepared by the direct mixing of the polyol, diol and diisocyanate, with the polyol and diol mixed before the reaction with MDI:

$$H{\left[OCH_2-CH_2-CH_2-CH_2-O\overset{O}{\overset{\|}{C}}-CH_2-CH_2-CH_2-CH_2-\overset{O}{\overset{\|}{C}}\right]}_b$$

Hydroxyl polybutylene adipate

$$+\ HOCH_2-CH_2-CH_2-CH_2-CH_2-OH\ +\ OCN-\!\!\bigcirc\!\!-CH_2-\!\!\bigcirc\!\!-NCO$$

$$OCH_2-CH_2-CH_2-CH_2-OH$$

1, 4–butanediol

MDI

$${\left\{{\left[O(CH_2)_4\,\overset{O}{\overset{\|}{O C}}(CH_2)_4\,\overset{O}{\overset{\|}{C}}\right]}_b\,O(CH_2)_4-O{\left[\overset{O}{\overset{\|}{C}}NH-\!\!\bigcirc\!\!-CH_2\right.}\right.}$$

$$\bigcirc\!\!-NH-CO\,(CH_2)_4\,O{\Big]}_a\,\overset{O}{\overset{\|}{C}}NH-\!\!\bigcirc\!\!-CH_2-\!\!\bigcirc\!\!-NH\overset{O}{\overset{\|}{C}}-{\Big\}}$$

Block copolymer

Poly (butylene adipate)/butanediol/MDI polyurethane ${\left[a-b\right]}_x$

9.3.2 Copolymerization Processes

These are batch processes, operating on low annual capacities, in the range of 5000 t/year. One example is the production of polyester urethane elastomer.

The polyester is prepared in two parallel lines. The operating time of 18 h includes the following steps: introduction of 1,4-butanediol and catalyst into the esterification reactor, vitrified and jacketed and equipped with a condenser. The butanediol is added to the reactor, and then heated to 140°C. The butanediol/adipic acid mole ratio is approximately 10. The reaction, which takes place at about 200°C, lasts 10 h.

The reaction mixture is then transferred to the polycondensation reactor. The reaction time is 15 h. The excess butanediol is distilled and sent to the rectification section. The polyester formed ($\overline{M_n} \approx 1000$) is cooled to 100°C and pumped to a storage tank. After mixing with 1,4-butanediol and catalyst, the polyester is sent to the double screw extruder which serves as a polymerization reactor. At the exit and after cooling, the extrudates, reduced to pellets, are sent to a storage tank.

9.3.3 Economic Data

Table 9.5 provides a glance at the economic data for a capacity of 5000 t/year of polyester urethane elastomer. This low unit capacity corresponds to the average capacities observed in the United States and Western Europe.

TABLE 9.5

POLYESTER URETHANE THERMOPLASTIC RUBBER

Capacity (t/year)	5000
Battery limits investment: 10^6 FF in 1989	80
Consumption per ton of thermoplastic rubber:	
· Raw materials:	
MDI (t)	0.41
1,4-butanediol (t)	0.37
Adipic acid (t)	0.38
· Utilities:	
Steam (t)	6
Cooling water (m^3)	120
Electricity (kWh)	600
Labor (operators per shift)	6

9.4 COPOLYESTER/ETHER THERMOPLASTIC ELASTOMERS

9.4.1 General

These products were developed and marketed in the early 1970s by *Du Pont de Nemours* under the brand name Hytrel. Their two-phase structure consists of rigid crystalline blocks of polybutylene terephthalate (polyester) and rubbery amorphous blocks of polytetramethylene ether glycol terephthalate (polyether):

Crystalline block   Amorphous block

The respective proportions of the two crystalline and amorphous blocks have a decisive influence on the physicochemical and mechanical properties of these products. The rigid crystalline polyester blocks tend to strengthen the polymer. The amorphous polyether blocks give rise to the elastomer property observed. The mechanical properties are similar to those of vulcanized elastomers: in particular, they display good tensile and tear strength. These products show good resistance to oils, chemicals and abrasion. The maximum service temperatures range between -36 and $+150°C$.

9.4.2 Polymerization Process

The process is a batch process in two steps. The polymer is formed from dimethylterephthalate (DMT), poly(tetramethylene) glycol (PTMEG) and 1,4-butanediol (1,4-BD).

The first or pre-polymerization step is an exchange reaction in which 1,4-BD and PTMEG are substituted for the ethyl groups of DMT. This transesterification is conducted at atmospheric pressure between 130 and 150°C. The catalyst, a derivative of tetrabutyl titanate, is dispersed in the 1,4-BD, and then added to the reaction mixture. The stainless steel reactor (316 SS) is equipped with a reflux column. The reaction is controlled by the quantity of methanol leaving the top of the reactor. The methanol begins to distill from 165°C. The reaction stops when the temperature reaches 200°C. It is complete in 1 h.

The pre-polymer is then transferred under vacuum to the polymerization reactor, which makes up the second step. The jacketed stainless steel reactor is equipped with an anchor stirrer. The reaction is conducted under vacuum at 250°C. Polycondensation time is 2 h. The 1,4-BD released is condensed, pumped to the purification column, and then recycled to the reactor. When the polymer reaches the desired intrinsic viscosity, it is discharged from the reactor under inert atmosphere, and transferred to an extruder. The polymer strands at the extruder outlet are cooled with water, cut into pellets, and dried in a rotary drum drier.

Table 9.6 offers some economic data on the production of a copolyester/ether for a unit capacity of 5000 t/year representative of the average installed capacities.

TABLE 9.6

COPOLYESTER/ETHER THERMOPLASTIC RUBBER

Capacity (t/year)	5000
Battery limits investment:	
10^6 FF in 1989	60
Consumption per ton of thermoplastic rubber:	
· Raw materials:	
Dimethylterephthalate (t)	0.595
Polytetramethylene ether glycol terephthalate (t)	0.380
1,4-butanediol (t)	0.250
· Catalysts and chemicals (FF):	20
· Co-product: methanol (t)	− 0.18
· Utilities:	
Steam (t)	0.8
Process water (m^3)	20
Cooling water (m^3)	40
Electricity (kWh)	360
Labor (operators per shift)	5

9.5 OTHER ELASTOMERS

Three main types of elastomer are in full development today.

9.5.1 Interpenetrating Polymer Network (IPN)

Marketed by *Shell*, this block copolymer helps to favor blends of incompatible polymers or to form alloys of polymers. The Kraton® series are elastomers consisting of styrene, ethylene/butene, styrene blocks mixed with a thermoplastic resin, with the combination forming an IPN network.

9.5.2 Polyether/Polyamide Block Copolymer

Atochem introduced this new product on the market under the brand name Pebax. The structure consists of rigid blocks of polyamides (nylon 11 or 12) and flexible polyether blocks (polyoxyethylene, polyoxypropylene).

9.5.3 Styrene, Ethylene/Butene, Styrene Block Copolymer Modified with a Polysiloxane

These products, marketed by *Concept Products* (USA) under the trademark name C-Flex, are polymers that can compete with PVC, silicon rubbers and polyurethanes.

The polysiloxane is generally polydimethylsiloxane. This copolymer differs from SBC in the much greater frequency of alternating rigid and flexible blocks in the chain.

9.6 USES AND PRODUCERS

The main branches of application of these new elastomers are the following:

(a) Adhesives: the main uses of styrene/diene copolymers are adhesive tapes obtained from hot melts or adhesives in solution, adhesives for paper or cardboard stripping, contact adhesives and adhesives for the building trade (SBC).
(b) Shoes: this includes compact or microcellular soles, thick soles, injection molding of boot heels, ski boots (TPU, COPE).
(c) Automotive: this is potentially the largest sector, but still not fully developed, in which current applications concern bumpers, hoses, electrical insulation (SBC), rack and pinion steering gear boots, constant velocity joint boot seals to protect front wheel drive automobile mechanisms (COPE), injection molded automotive components, car front ends (TPU).
(d) Protective coatings applied to miscellaneous leather articles or metallic parts, electrical wires and cables (SBC for heat- and age-resistant electrical insulation applications).
(e) Hoses and tubes: high-pressure hydraulic hoses for brake systems of trucks (COPE), hydraulic hose (TPU), geophysical cable jacketing used in oil exploration (TPU).
(f) Blend with asphalt, especially for the building sector, to improve the temperature resistance of asphalt: single-ply roofing, road repair and paving materials (SBC).
(g) Viscosity index improvers: styrene/hydrogenated isoprene block copolymers used by *Shell* in the formulation of multi-grade oils (Shellvis®).
(h) Blends with thermoplastics to reinforce the impact and tensile tear strength of polyethylene, polypropylene, polystyrene and engineering thermoplastics (SBC).
(i) Consumer articles: molded and extruded products necessary for consumer goods such as toys, sports equipment, household wares (SBC).

Although they are most costly than other elastomers, interest is growing in thermoplastic elastomers due to:

(a) Less energy consumption for their manufacture.
(b) Better lightness, rigidity and stability.
(c) Possibility of adhesive formulas by reducing the amount of organic solvents required.

The production capacities of the different thermoplastic rubbers in 1989 are given in Table 9.7.

TABLE 9.7
PRODUCTION CAPACITIES OF THERMOPLASTIC ELASTOMERS IN 1989
(IN THOUSANDS OF TONS PER YEAR)

Country	Styrene-based block copolymers	Polyurethanes	Copolyesters/ ethers
USA	140	30*	40
Japan	55	10*	6
Italy	65		
Belgium	45		
Spain	40	50*	
France	35		
West Germany	35		
Luxembourg			20
World	**415**	**90***	**66**

* Estimated from production and consumption figures.

APPENDIXES

Appendix 1
Comparative properties of synthetic rubbers and natural rubber

	Natural rubber	SBR	Poly-buta-diene	Poly-Iso-prene	Ethylene/propy-lene rubber	Butyl rubber	Poly-chloro-prene rubber	Nitrile rubber
Dielectric properties	VG	G	G	VG	G	VG	M	M
Tensile strength	E	G	G	E	G	G	VG	G
Tear strength	VG	G	VG	VG	VG	M	G	G
Abrasion resistance	E	VG	E	E	VG	G	G	G
Resistance to repeated bending	VG	G	VG	VG	G	G	VG	G
Resilience	E	G	E	E	G	M	VG	G
Creep resistance	G	G	G	G	VG	M	G	G
Ageing resistance	M	G	G	M	E	VG	VG	G
Resistance to ozone	N	N	N	N	E	VG	VG	N
Resistance to heat	M	G	M	M	VG	G	G	G
Resistance to cold	VG	G	VG	VG	VG	G	M	M
Resistance to aliphatic hydrocarbons	N	N	N	N	N	N	G	VG
Resistance to aromatic hydrocarbons	M	M	M	M	G	G	M	N
Resistance to ketone solvents	N	N	N	N	N	M	M	M
Resistance to chlorinated solvents	N	N	N	N	N	M	M	M
Resistance to water	G	G	G	G	VG	VG	G	G
Resistance to dilute acids	G	G	G	G	E	E	VG	G
Resistance to strong acids	M	M	M	M	VG	VG	G	M
Resistance to strong oxidizing acids	N	N	N	N	M	M	N	N
Impermeability to gases	M	M	M	M	M	E	G	G

Key: E Excellent   VG Very good   G Good
      M Mediocre   N None

J.P. ARLIE

Appendix 2
Principal brands and producers of synthetic elastomers

The following acronyms are routinely used to denote synthetic elastomers:

SBR	Styrene/butadiene rubber
BR	Polybutadiene
IR	Polyisoprene
EP or EPDM	Ethylene/propylene rubber
IIR	Butyl rubber
CR	Polychloroprene
NBR	Nitrile rubber
TPE	Thermoplastic elastomers
SBC	Styrene block copolymers
TPU	Polyurethane block copolymers
COPE	Polyester block copolymers
PBA	Polyether block amide
IPN	Styrene rubber/thermoplastic resin blends

Brand	Type of elastomers	Producer
AMERPOL	SBR	Goodrich (USA)
AMERIPOL CB	BR	Goodrich (USA)
AMERIPOL SN	IR	Goodrich (USA)
ARLASTIC	TPE/COPE	AKZO (Netherlands)
ARNITEL	TPE/COPE	AKZO (Netherlands)
ASAPRENE	TPE/COPE	Japan Elastomer Company
ASRC	SBR	American Synthetic Rubber Corporation (USA)
BAYPREN	CR	Bayer (West Germany)
BAYTOWN	SBR	American Synthetic Rubber Corporation (USA)
BREON	NBR	British Petroleum (United Kingdom)
BUDENE	BR	Goodyear (USA)
BUNA OP, S, SS	SBR	Hüls (West Germany)
BUNA CB	BR	Hüls (West Germany)

APPENDIXES

Brand	Type of elastomers	Producer
BUNA NB 188	NBR (+ poly-vinyl chloride)	*Hüls* (West Germany)
BUTACRIL	NBR	*PCUK* (France)
BUTACLOR	CR	*Rhône-Poulenc* (France)
BUTAPRENE	SBR or NBR	*Firestone* (USA)
CALPRENE	TPE/SBC	*REPSOL* (Spain)
CARIFLEX BR	BR	*Shell*
CARIFLEX S	SBR	*Shell*
CARIFLEX TR	TPE/SBC	*Shell*
CHEMMIGUM	NBR	*Goodyear*
CIS-4	BR	*Phillips Petroleum* (USA)
CISDENE	BR	*American Synthetic Rubber Corporation* (USA)
COPO, CARBOMIX	SBR	*Copolymer Rubber and Chemical Corporation* (USA)
CYANAPRENE	TPE/TPU	*American Cyanamid Company* (USA)
DENKA CHLOROPREN	CR	*Denki Kagaku* (Japan)
DIENE	BR	*Firestone* (USA)
DURAGEN	BR	*General Tire and Rubber Company* (USA)
DUTRAL CO, TR	EP	*Montedison* (Italy)
ECDEL	TPE/COPE	*Eastman Kodak* (USA)
EKAPRIM	NBR	*Montedison* (Italy)
EPCAR	EP	*Goodrich*
EPSYN	EP	*Copolymer Rubber and Chemical Corporation* (USA)
ESTANE	TPE/TPU	*Goodrich*
EUROPRENE N	NBR	*Anic* (Italy)
EUROPRENE SOL T	TPE.SBC	*Anic* (Italy)
EUROPRENE CIS	BR	*Anic* (Italy)
FINAPRENE	TPE/SBC	*Finaprene* (Belgium)
FR-N	NBR	*Firestone* (USA)
FR-S	SBR	*Firestone* (USA)

Brand	Type of elastomers	Producer
GENTRO and GENTRO JET	SBR	*General Tire and Rubber Company* (USA)
HYCAR	NBR	*Goodrich*
HYTREL	TPE/COPE	*Du Pont de Nemours* (France)
INTENE	BR	*International Synthetic Rubber* (United Kingdom)
INTOL	SBR	*International Synthetic Rubber* (United Kingdom)
JSR	SBR, BR, NBR	*Japan Synthetic Rubber*
KELTAN	EP	*DSM* (Netherlands)
KRATON	TPE/SBC	*Shell*
KRYLENE	SBR	*Polysar International* (Switzerland)
KRYNAC	NBR	*Polysar International* (Switzerland)
KRYNOL	SBR	*Polysar International* (Switzerland)
NAIRIT	CR	USSR
NATSYN	IR	*Goodyear*
NAUGAPOL	SBR	*Uniroyal* (USA)
NEOPRENE	CR	*Du Pont de Nemours* (France)
NIPOL	BR	*Japanese Geon* (Japan)
NY SYN	NBR	*Copolymer Rubber and Chemical Corporation* (USA)
PARACRIL	NBR	*Uniroyal* (USA)
PELLETHANE	TPE/TPU	*Upjohn* (USA)
PERBUNAN N	NBR	*Bayer* (West Germany)
PLIOFLEX	SBR	*Goodyear*
PLIOFLEX 5000	BR	*Goodyear*
POLYSAR S	SBR	*Polysar International* (Switzerland)
PRO-FAX	TPE/TPU	*Hercules* (USA)
ROYALAR	TPE/TPU	*Uniroyal* (USA)
ROYALENE	EP	*Uniroyal* (USA)
SIRBAN	NBR	*SIR* (Italy)

Brand	Type of elastomers	Producer
SKS	SBR	USSR
SKN	NBR	USSR
SKB, SKD, SKV	BR	USSR
SKI, SKI.3	IR	USSR
SKYPRENE	NBR	*Toyo Soda* (Japan)
SOLPRENE	SBR	*Firestone* (USA)
SOMEL	TPE	*Du Pont de Nemours* (France)
STEREON	SBR	*Firestone* (USA)
SYNPOL	SBR	*Texas US Chemical Company* (USA)
TAKELAC	TPE/TPU	*Takeda Chemical* (Japan)
TAKTENE	BR	*Polysar International* (Switzerland)
TELCAR	TPE/TPU	*Goodrich*
TEXIN	TPU	*Mobay Chemical Corporation* (USA)
TPR	TPE	*Uniroyal* (USA)
TUFPRENE	TPE/SBC	*Asahi Chemical* (Japan)
TUFTEC	TPE/SBC	*Asahi Chemical* (Japan)
UGITEX	BR	*PCUK* (France)
UNEPRENE	TPE	*International Synthetic Rubber* (United Kingdom)
VISTAFLEX	TPE/TPU	*Exxon* (USA)
VISTAFLON	EP	*Exxon* (USA)

REFERENCES

GENERAL WORKS

Encyclopedia of Polymer Science and Technology, Interscience Publishers, New York, 1971.

Harper, C.A., *Handbook of Plastics and Elastomers*, McGraw-Hill, New York, 1975.

Kennedy, J.P., Tornqvist, E.G.M., *Polymer Chemistry of Synthetic Elastomers*, Interscience Publishers, New York, 1968.

Kirk-Othmer, *Encyclopedia of Chemical Technology*, Interscience publishers, New York, 1971.

MacKetta, *Encyclopedia of Chemical Processing and Design*, Marcel Dekker Inc., New York, 1976.

Morton, M., *Rubber Technology* (2nd edition), Van Nostrand Reinhold Company, New York, 1973.

Ritchie, P.D., *Vinyl and Allied Polymers*, **1**, Plastics Institute, London, 1968.

Winding, G., Hiatt, G., *Plastics Materials*, McGraw-Hill, 1961.

Witby, G.S., *Synthetic Rubber*, Wiley, New York, 1957.

PAPERS CONCERNING A SPECIFIC SUBJECT

Aboulafia, J., *Revue de l'Institut Français du Pétrole*, **25**, No. 4, 1970, p. 472.

Back, A.L., "Solution polymerization", *Chemical Engineering*, Aug. 1966, p. 65.

Baldwin, F.P., Verstrate, G., "Polyolefin elastomer based on ethylene and propylene", *Rubber Chemistry and Technology*, **45**, 1972, p. 709.

Bawn, C.E.H., "Polymerization of butadiene by soluble Ziegler Natta catalyst", *Rubber and Plastics Age*, **46**, 1965, p. 510.

Beatty, J.R. *et al.*, "Butadiene styrene rubbers", *Industrial and Engineering Chemistry*, **44**, 1952, p. 742.

Binder, J.L., "Microstructures of polybutadienes and butadiene styrene copolymer", *Industrial and Engineering Chemistry*, **46**, 1954, p. 1727.

Caoutchouc Nitrile Butacril®, notice technique, Plastimer-Ugine Kuhlmann, 1976.

Caoutchouc Polychloroprene Butaclor®, notice technique, Distugil, 1979.

REFERENCES

Crespi, G., Di Drusco, G., "Make EPDM by suspension process", *Hydrocarbon Processing*, 1979, p. 102.

Dumon, R., "L'avenir des caoutchoucs synthétiques", *Information Chimie*, **191**, 1979, p. 140.

Dworkin, D., "Changing markets and technology for specialty elastomers", *Rubber World*, 1975, 43.

Francis, D.H. *et al.*, "GRS continuous polymerization: factors and equipment used in process", *Chemical Engineering Progress*, **45**, 1949, p. 102.

Gippin, M., "Polymerization of butadiene with alkyl aluminium compounds and cobalt chloride", *Industrial and Engineering Chemistry, Process Design and Development*, **1**, 1962, p. 32, **3**, 1965, p. 160.

Hoffmann, W., "Nitrile rubber", *Rubber Chemistry and Technology*, **37**, No. 2, 1964.

Hsieh, H.L., "Solution polymerization initiated with alkyl lithium", *Rubber and Plastics Age*, **46**, 1965, p. 394.

Kuntz, I. *et al.*, "Butyl lithium initiated polymerization of 1,3 butadiene", *Journal of Polymer Science*, **42**, 1960, p. 299.

Johnson, P.R., "Polychloroprene rubber", *Rubber Chemistry and Technology*, **49**, No. 3, 1976.

Miller, S.A., "Synthetic rubber", *Chemical and Process Engineering*, 1971, p. 57.

Prince, A.K. *et al.*, "Synthetic rubber production", *Industrial and Engineering Chemistry*, **52**, No. 3, 1960, p. 235.

Schlegel, W.F., "Design and scaleup of polymerization reactors", *Chemical Engineering*, March 1972, p. 88.

Schoenberg, E. *et al.*, "Polyisoprene", *Rubber Chemistry and Technology*, **52**, 1979, p. 526.

Scroeder, E.E., "Recent developments in synthetic rubber technology", *Chemical Engineering*, Nov. 1966, p. 132.

INDEX

J.P. ARLIE

ACHEVÉ D'IMPRIMER
SUR LES PRESSES DE
L'IMPRIMERIE CHIRAT
42540 ST-JUST-LA-PENDUE
EN MAI 1992
DÉPÔT LÉGAL 1992 N° 6908
N° D'ÉDITEUR 849

IMPRIMÉ EN FRANCE

—